Series in BioEngineering

The Series in Bioengineering serves as an information source for a professional audience in science and technology as well as for advanced students. It covers all applications of the physical sciences and technology to medicine and the life sciences. Its scope ranges from bioengineering, biomedical and clinical engineering to biophysics, biomechanics, biomaterials, and bioinformatics.

Indexed by WTI Frankfurt eG, zbMATH.

Ab Aziz Mohd Yusof · Muhamad Noor Harun ·
Ardiyansyah Syahrom

Rowing Biomechanics and Hydrodynamics

Performance Enhancement Through Sport Engineering

Springer

Ab Aziz Mohd Yusof
School of Mechanical Engineering
College of Engineering
Universiti Teknologi MARA
Masai, Johor, Malaysia

Muhamad Noor Harun
Faculty of Mechanical Engineering
Universiti Teknologi Malaysia
Skudai, Johor, Malaysia

Ardiyansyah Syahrom
Department of Applied Mechanics
and Design
Faculty of Mechanical Engineering
Universiti Teknologi Malaysia
Skudai, Johor, Malaysia

ISSN 2196-8861 ISSN 2196-887X (electronic)
Series in BioEngineering
ISBN 978-981-96-1871-2 ISBN 978-981-97-7271-1 (eBook)
https://doi.org/10.1007/978-981-97-7271-1

© The Editor(s) (if applicable) and The Author(s), under exclusive license to Springer Nature Singapore Pte Ltd. 2025

This work is subject to copyright. All rights are solely and exclusively licensed by the Publisher, whether the whole or part of the material is concerned, specifically the rights of translation, reprinting, reuse of illustrations, recitation, broadcasting, reproduction on microfilms or in any other physical way, and transmission or information storage and retrieval, electronic adaptation, computer software, or by similar or dissimilar methodology now known or hereafter developed.
The use of general descriptive names, registered names, trademarks, service marks, etc. in this publication does not imply, even in the absence of a specific statement, that such names are exempt from the relevant protective laws and regulations and therefore free for general use.
The publisher, the authors and the editors are safe to assume that the advice and information in this book are believed to be true and accurate at the date of publication. Neither the publisher nor the authors or the editors give a warranty, expressed or implied, with respect to the material contained herein or for any errors or omissions that may have been made. The publisher remains neutral with regard to jurisdictional claims in published maps and institutional affiliations.

This Springer imprint is published by the registered company Springer Nature Singapore Pte Ltd.
The registered company address is: 152 Beach Road, #21-01/04 Gateway East, Singapore 189721, Singapore

If disposing of this product, please recycle the paper.

Preface

Rower biomechanics, stroke style, and hydrodynamics of the blade are among the essential factors influencing rowing performance. A deeper understanding of these aspects will help the rower and rowing fraternity decide on the proper technique and equipment for better performance. Previous researchers have explored the forces profile, physiological and anthropometry of the rower, fluid dynamic around the blade, and blade design optimisation. However, the coupling mechanism or incorporation between the biomechanics of the rower and the hydrodynamic received less attention.

The book's main objective is to introduce and explore the biomechanics and hydrodynamics of rowing. It also aims to introduce engineering in sports and its application toward performance enhancement. The targeted reader is an academician in the rowing sports community and sports researcher either for understanding the sport from an engineering perspective or for the performance. Generally, the book is divided into seven chapters covering rowing sport, biomechanics, and related hydrodynamics.

Chapter 1 provided a basic overview of rowing. The history of rowing was introduced in this chapter, and the structural contents focused on rowing-related governing organisations to set rules and regulations related to race events. The four-stroke phases were introduced in this chapter with their illustration. The engineering aspect, such as the free-body diagram of the moving boat during the competition, was presented. The force acting on the blade elaborated more since the propulsive force used to drive the boat was generated when the blade was swept during the drive phase.

Chapter 2 deals with the biomechanical and hydrodynamic of rowing. The related aspects were highlighted, such as the biomechanics of rowing and the contribution of the body segments and stroke styles. Rower performance can be monitored based on the force profile. Important parameters such as maximum peak force, percentage of drive to achieve the peak, stroke rate, and total rotated angle were extracted from the previous study in a systematic review. The important elements of the hydrodynamic force generated on the blade were presented. Basic hydrodynamic formulas were included in this section. The magnitude of hydrodynamic force was influenced by surface area, blade coefficient, and relative speed. The force is then resolved as components based on the boat's direction and blade axis for further analysis. An example of force calculation based on the experimental data from the previous study was provided. In the calculation, the propulsive force generated on the blade resolved into an axis component and managed according to the force vector. Important engineering aspects can be applied to the blade parameter and its design as an innovation aspect, leading to performance enhancement.

Chapter 3 highlights the need to incorporate biomechanics and hydrodynamics in a study. This suggested method and analysis bring the rowing to significant progress in the biomechanics and hydrodynamic aspects. The book tries to incorporate biomechanism and hydrodynamics as a comprehensive study on understanding oar, which is important. Biomechanics was applied on the handle, and hydrodynamic force was generated on the blade. The study can be simplified based on several assumptions, such as controlled motion, simplified kinematics, and less interaction between rowers.

In Chap. 4, as the method to study hydrodynamics further develops, the use of computational fluid dynamics can be considered. The hydrodynamic of the rowing is accomplished using the finite volume method, which is governed by the incompressible Navier-Stokes equation. Several examples of previous studies and investigations of necessary model simplification were highlighted in this book for the reader's convenience.

The important aspect of pressure and force generated due to the interaction between the Oar blade and water was discussed in Chap. 5. Studying fluid dynamics in rowing is crucial for improving performance and understanding the sport's physics. Fluid-structure interaction (FSI) methods, combining computational fluid dynamics (CFD) with finite element analysis (FEA), are used to simulate how fluid interacts with the oars. An analytical method for calculating FSI was developed to investigate blade design parameters and their impact on rowing performance. Understanding the interplay between fluid flow, solid structure mechanics, and athlete abilities can lead to better hydrodynamic propulsion, especially in rowing.

Chapter 6 employed computational simulation methods, utilizing Star CCM software, to analyse the hydrodynamic forces involved in rowing propulsive mechanics under unsteady conditions of incompressible turbulence flow. The numerical simulations focused on the flow around the blade during the drive phase and recorded the corresponding hydrodynamic forces. Quasi-static and dynamic blade movement due to the rower is being focused. Validation was conducted in the quasi-static study for a rectangular blade based on previous experiments involving quarter-sized blade models. Meanwhile, for the dynamic study, the blade was rotated with specific angular velocity to replicate the drive phase of the rowing stroke. The dynamic study employed the coupled mesh grids at the overlapping interfaces, and interpolation was used at these interfaces to obtain fluid variables, allowing for a comprehensive study of dynamic blade rotation due to rower force profiles.

Chapter 7 discusses the experimental aspect of the rowing study. To address this gap, a study was conducted to examine the coupling mechanism between rower biomechanics, specifically kinematics and handle force, and the hydrodynamics of the blade during the drive phase. A dynamic rowing simulator equipped with specialized software was employed for data collection, aiming to replicate the on-water rowing experience indoors. Sensors attached to the simulator gathered crucial data to track and enhance athletes' performance. This data included stroke rate, which helps maintain optimal pacing, force curves revealing force application patterns, and power generation metrics, offering valuable insights for performance assessment and improvement.

Johor, Malaysia

Ab Aziz Mohd Yusof
abaziz86@uitm.edu.my

Muhamad Noor Harun
mnoorharun@utm.my

Ardiyansyah Syahrom
ardi@utm.my

Acknowledgements

First and foremost, praise and thank God for the opportunity, physical and mental strength, and valuable time to complete this book, *Rowing Biomechanics and Hydrodynamics: Performance Enhancement Through Sport Engineering*. A special dedication of appreciation is given to all fellow members for help, guidance, ideas, and advice through the data collection, editing, and book preparation. We want to thank PERDAMA for its trustworthiness and the opportunity to use the rowing facilities and equipment to achieve the book's objectives. We highly acknowledge the association's expertise in rowing. We also would like to express our gratitude to all MEDITEG, SITC, UTM, and UiTM members for their assistance over the last few years. Special dedication to Hazuwa, Adhwa, and Adira for the unending inspiration in this journey.

Contents

About the Authors

Dr. Ab Aziz Mohd Yusof received his bachelor's degree in Mechanical engineering from Universiti Teknologi Malaysia, UTM (Malaysia), in 2010 and Ph.D. in Biomedical engineering in 2017 from Universiti Teknologi Malaysia, UTM (Malaysia). He is a senior lecturer at the Faculty of Mechanical Engineering, UiTM Pasir Gudang. His research interests are computational study, sports engineering, and biomechanics.

Assoc. Prof. Dr. Muhamad Noor Harun received his bachelor's degree in Mechanical Engineering at Universiti Teknologi MARA in 2000. He continued his postgraduate (Ph.D.) studies in Mechanical engineering at Loughborough University, UK, and received a Ph.D. in 2008. He is a senior lecturer at the Faculty of Mechanical Engineering and a member of the Sports Innovation and Technology Center (SITC). He is interested in tribology and sports engineering.

Assoc. Prof. Dr. Ardiyansyah Syahrom graduated with a first degree from the Universitas Bung Hatta in 2000. After serving Universiti Teknologi Malaysia as a research officer for a year, he entered the master's in Mechanical Engineering from Universiti Teknologi Malaysia and graduated in 2006. He then pursued his study at Universiti Teknologi Malaysia, Malaysia, from 2007 to 2011 and was also awarded a Ph.D. degree from this University. He is actively involved in Medical Devices, Biomechanics, Biomaterial, Bone, and Sports Engineering.

List of Figures

List of Tables

Chapter 1
Introduction and Basic Principles of Rowing

1.1 Rowing Background

Rowing is one of the oldest Olympic sports and has been a part of the Olympic event since 1900 [1, 2]. Rowing also competed in the World Rowing Championships, a week-long event held in non-Olympic years [3]. The World rowing championships had been discussed at the international level several times before the event started in 1962 and then were held annually from 1972 onwards. This sport was also introduced in the Commonwealth Games as an optional sport. The first record of Commonwealth games was conducted in 1911, known as the Festival of Empire, while the first rowing game was introduced in 1930 [4].

Most of the event is organised by the International Federation of Rowing Associations, FISA, the oldest international sports federation in the Olympic movement. FISA is a governing organisation for the sport of rowing [5]. The function of FISA is to set the rules and regulations for rowing competitions, with the service of advice and expertise to organise rowing regatta and coaching education.

Rowing events are subjected to rules and regulations to provide safety for competitors, ensure fair race, and promote the consistent application of rules [6]. Generally, the race event has to follow the start signal from the starter without a false start or failure. Furthermore, each boat is assigned a and must not interface with others, or the referee shall instruct of offending. Finally, the boats must finish the race when the bow touches the finish line.

Rowing equipment can be divided into three essential parts: oar, rigger, and boat [7–10]. An oar is used as a lever to propel the boat. The rigger is placed at the side of the seat to support the pivot that holds the oar. The rowing boat consists of a sliding seat, stern, and shoes.

The rowing boat is propelled using the oar where the blade is submerged, and the handle is pulled to generate the hydrodynamic force. The performance depends on the use of human ability and sports equipment. Both aspects should match adequately to compete successfully in the highest-level competitions.

© The Author(s), under exclusive license to Springer Nature Singapore Pte Ltd. 2025

A. A. Mohd Yusof et al., *Rowing Biomechanics and Hydrodynamics*,
Series in BioEngineering, https://doi.org/10.1007/978-981-97-7271-1_1

A clear understanding of rower biomechanics is very well related to the rower's kinetic, kinematic, physiological, and anthropometrics, bringing significant attention to the sports community. Altering the function and utilisation of the body segment is required to generate proper biomechanics force to cope with the demand for sports performance.

The rowing stroke involves the coordination of legs, trunk, and arm segments at specific sequences before it is repeated. Therefore, this rowing action is categorised as a cyclic type of sport [11, 12]. The repeated stroke is divided into four phases: the catch, drive, finish, and recovery [12, 13]. Power delivery from rower biomechanics accelerates the boat during the drive phase.

Different styles of legs, trunks, and arms produce different rowing styles. It involves the coordination of the body position of the catch, timing, and body segment emphasis [14]. Rowing style and boat performance can be evaluated using the handle force profile generated during the stroke. Researchers have begun to explore the definition of the ideal force profile to enhance performance and maximise the utilisation of rower biomechanics [13]. Usually, the highest work generated by the rower produces the fastest boat speeds. However, as there is a limitation on the physiology of the rower's body, the most suitable rowing style needs in-depth exploration.

The hydrodynamics force is used to overcome the boat's drag and accelerate the boat. Skilful rowing enhances the hydrodynamic force and improves the boat's speed. In addition, using advanced technology to produce better equipment accelerates the enhancement process. The competition has three commonly used blades: Macon, Big, and Fat [15–17]. Thus, properly understanding the fluid flow around the blade is necessary to improve the blade's performance. Meanwhile, the blade slips problem, and the inefficiency of blade propulsive are among the main issues affecting rowing performance [18, 19]. Further assessment of each blade's features towards the hydrodynamic force deepens the understanding of the fluid mechanics of the blade propulsive.

Selection of the best rowing style and blade design is essential to increase rowing performance. Varieties of blade design and rowing styles used by the rower contribute to various rower kinematic and hydrodynamic performance of the blade. Rower kinematics can be assessed using video analysis where interested body segments are emphasised. Meanwhile, the hydrodynamic force of the blade-associated fluid flows around the blade of the different rowing styles can be extracted using computational study due to the limitations of the experimental research.

1.2 Rowing Racing Strategies

Rowing racing strategies are racing tactics that aim to maximise the boat's speed and efficiency. The race can be divided into three main parts: (1) start, (2) body and (3) finish [20–22]. For the 2000 m sprint race competition, the start is at the first 500 m, the race's body is at 1000 m, and the finish is at the last 500 m [23]. There are differences in the stroke rate and sweep distance [23–25]. As the race begins, the rower starts the row, known as "the start," during the start, the rower rows with precision strokes and produces high horsepower to get the boat speeds as quickly as possible.

The start is simple, but the technique has variances such as 1/4, 1/2, 3/4, and full stroke start, which are defined as the applied stroke distance per full stroke distance [20]. After several start strokes, the rower usually changes the stroke rhythm by increasing the stroke rate to a higher level or lengthening the stroke. The rower with good strokes and gains advanced speed at the start usually moves far forward and is comfortable manipulating the rest of the race. During the race, 65–75% of metabolic energy is produced [26]. At the same time, the rower lowers the stroke rate to cope with the body's aerobic metabolism and causes the boat to speed slower [27, 28]. The finished part is where the race outcome is decided. The rower increases the stroke rate and speeds to finish the race [23]. Usually, the rower rows by bringing out the start rate rhythm using anaerobic metabolism, like a start, which produces 25–35% energy to accelerate the boat faster [22, 26].

1.3 Rowing Stroke Phase

Rowing is generally classified as a cyclic event sport with repeated four-stroke phases, as shown in Fig. 1.1. The phases begin with a catch, followed by the drive, finish, and end with recovery before it is repeated in the same sequence [12, 29–31]. However, it is essential to note that only the drive phase of the rowing stroke contributes to the generation of hydrodynamic propulsion force. The others only complement the cycle [17, 32, 33].

During the catch phase, the blade is initially submerged, and then it is followed by the drive phase, where the blade travels through the arc path. The finish phase is achieved when the blade reaches the maximum travelling arc and is prepared to be lifted. The final phase is recovery, where the blade feathers travel in the air and then return to the catch position. Featuring is a part of the rowing stroke where the rower turns the blade from a perpendicular position to the water to a parallel position to reduce the air drag. Rowers transfer biomechanics power during the drive to accelerate the boat via water support conditions [29, 34]. On the contrary, the boat loses its acceleration due to its drag during the recovery. Thus, to improve rowing performance, most rowers focus on maximising force generation of the drive [33, 34].

Specifically, the drive phase can be separated into six flow region sections [35]. The classification is based on the relative blade motion during the drive phase and the development of the instantaneous blade force. The first two sections are essential, attributing to a higher efficiency section, which generates a more significant propulsive force. Due to this aspect, most rowers tend to increase force rapidly to achieve the peak force early. Once the blades move from catch to an angle lower than 90°, the pressure increases rapidly by generating hydrodynamic force to provide thrust to the boat [35].

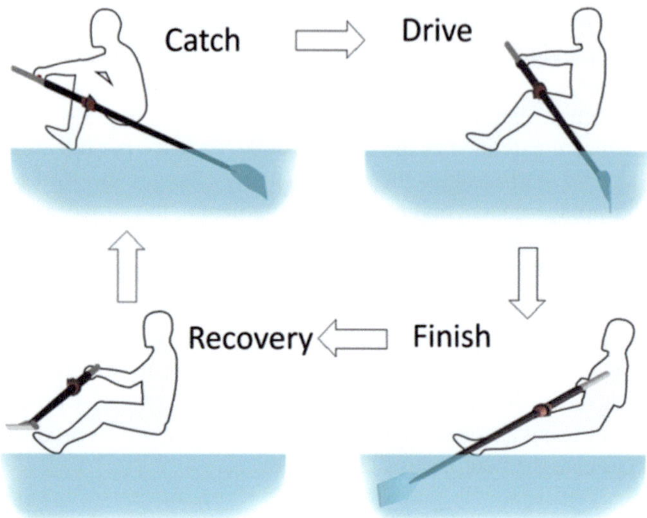

Fig. 1.1 Rowing stroke cycle

1.4 Free Body Diagram of Rowing

Figure 1.2 shows a free-body diagram of the rowing during the Varsity boat race 2016 held at Putrajaya. Multiple forces act on the rowing system. First, the rowing system is moved by the thrust force, F_T, which simultaneously functions to overcome the skin drag force on the boat F_D [36].

Using Newton's second law equation of motion of the boat and crew in Fig. 1.2 can be written as Eq. 1.2.1 [35]:

Fig. 1.2 Force and velocity act on the rowing system

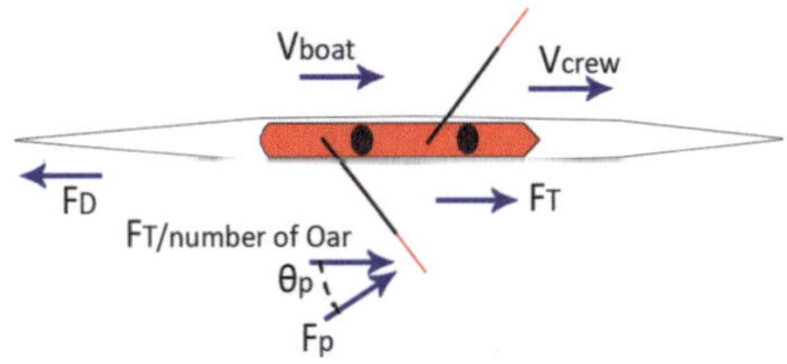

Fig. 1.3 The resultant hydrodynamic force acting during the stroke

$$F_T - F_D = m_{boat}\frac{dv_{boat}}{dt} + m_{crew}\left(\frac{dv_{crew}}{dt} + \frac{dv_{boat}}{dt}\right) \tag{1.2.1}$$

where F_T = boat trust force, F_D = skin drag force, m_{boat} = mass boat, and m_{crew} = mass crew.

The equation can be used to elucidate the role of thrust force in accelerating the boat while overcoming the skin drag resistance, F_D. Therefore, the boat's speed, V_{boat} is due to the applied thrust force.

As illustrated in Fig. 1.3, the thrust force is a component of the hydrodynamic force, which is resolved following the attacking angle of the blade as in Eq. 1.2.2.

$$\frac{F_T}{number\ of\ oar} = F_P \cos\theta_p \tag{1.2.2}$$

where F_P = Hydrodynamic force.

To understand the function of the propulsive force to the rowing motion, the blade is isolated, as in Fig. 1.4 [37].

The oar is rotated by pulling forces generated, as illustrated in Fig. 1.4, and it moves the boat in the transverse direction. Therefore, the propulsive force could be presented as the component of the boat's transverse motion and in the direction of the rotation motion of the blade, where FP is the resultant force of two force components. (Eq. 1.2.3).

$$F_P = F_x X_{Boat} + F_y Y_{Boat} = F_t h + F_n n \tag{1.2.3}$$

Meanwhile, F_x is the force that acts parallel to the direction of the moving boat, and F_y is the force that acts perpendicular to the direction of the moving boat. In contrast, Ft force acts along the length of the blade, and F_n force is perpendicular to the length of the blade.

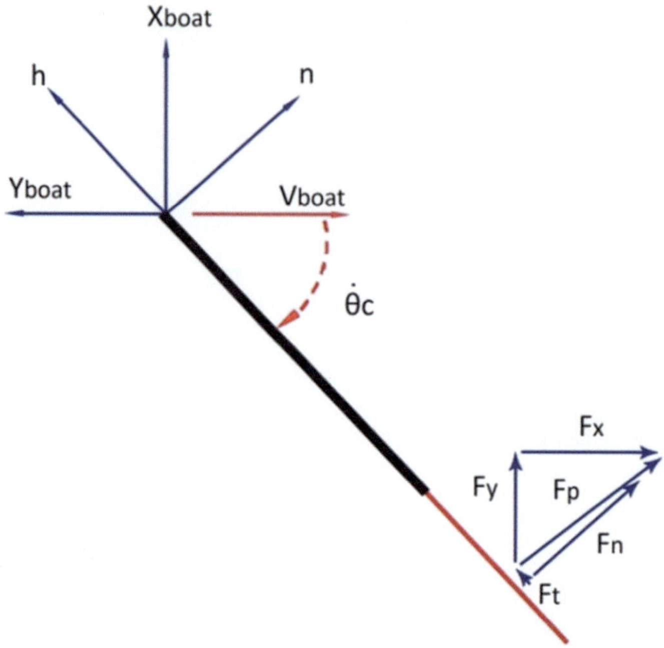

Fig. 1.4 Blade force components act on the rowing oar

References

1. De Coubertin, B.P.: The meeting of the Olympian games. North Am. Rev. **170**(523), 802–811 (1900)
2. Hernández, C.M.: History of Olympic rowing in Cuba. In: Proceedings: International Symposium for Olympic Research. International Centre for Olympic Studies (2006)
3. Steinacker, J.M., et al.: Training of rowers before world championships. Med. Sci. Sports Exerc. **30**(7), 1158–1163 (1998)
4. Perkin, H.: Teaching the nations how to play: sport and society in the British empire and Commonwealth. Int. J. Hist. Sport **6**(2), 145–155 (1989)
5. Sport, R., Paintworks, D.: FISA Rule Book
6. Committee, U.R.R.: Rules of Rowing. The United States Rowing Association, Princeton, NJ (2008)
7. Laschowski, B., et al.: The effects of oar-shaft stiffness and length on rowing biomechanics. Proc. Inst. Mech. Eng. Part P: J. Sports Eng. Technol. 1754337115582121 2015
8. Greidanus, A., Delfos, R., Westerweel, J.: Rowing faster by surface treatment. Procedia Eng. **34**, 361–366 (2012)
9. Smith, R.: The relationship between seat movement and boat acceleration during sculling. In: ISBS-Conference Proceedings Archive (1999)
10. Barrett, R.S., Manning, J.M.: Relationships between rigging set-up, anthropometry, physical capacity, rowing kinematics and rowing performance. Sports Biomech./Int. Soc. Biomech. Sports **3**(2), 221–235 (2004)
11. Kleshnev, V.: Segments contribution to the stroke length and rower power. Higher percentage of the trunk power helps to achieve higher boat speed. Rowing Biomech. Newslett. **2**(2) (2002)
12. Caplan, N., Coppel, A., Gardner, T.: A review of propulsive mechanisms in rowing. Proc. Inst. Mech. Eng. Part P: J. Sports Eng. Technol. **224**(1), 1–8 (2010)

13. Baudouin, A., Hawkins, D.: A biomechanical review of factors affecting rowing performance. Br. J. Sports Med. **36**(6), 396–402 (2002)
14. Kleshnev, V.: Rowing styles. Rowing Biomech. Newslett. **6** (60) (2006)
15. Caplan, N., Gardner, T.N.: A fluid dynamic investigation of the Big Blade and Macon oar blade designs in rowing propulsion. J. Sports Sci. **25**(6), 643–650 (2007)
16. Coppel, T.G.A., Caplan, N., Hargreaves, D.: Oar blade force coefficients and a mathematical model rowing. In: 27 International Conference on Biomechanics in Sports (2009)
17. Caplan, N., Gardner, T.N.: Optimization of oar blade design for improved performance in rowing. J. Sports Sci. **25**(13), 1471–1478 (2007)
18. Kleshnev, V.: Propulsive efficiency of rowing. In: Proceedings of the XVII International Symposium on Biomechanics in Sports (1999)
19. Affeld, K., Schichl, K., Ziemann, A.: Assessment of rowing efficiency. Int. J. Sports Med. **14**, S39-41 (1993)
20. Kleshnev, V.: Analysis of the race start. Rowing Biomech. Newslett. **9**(1) (2009)
21. Kleshnev, V.: Racing strategy in rowing during Sydney Olympic Games. Aust. Rowing **24**(1), 20–23 (2001)
22. Nilsen, T.S., Daigneault, T., Smith, M.: Basic Rowing Physiology. FISA Development Commission (1990)
23. Doyle, M.M., Lyttle, A., Elliott, B.: Comparison of force-related performance indicators between heavyweight and lightweight rowers. Sports Biomech. **9**(3), 178–192 (2010)
24. Kleshnev, V., Stroke rate vs. distance in rowing during the Sydney Olympics. Australian Institute of Sport Australian Rowing, 2001. 24(2), : p. 18–22.
25. Kleshnev, V.: The effect of stroke rate on biomechanical parameters and efficiency of rowing. In: XIV Symposium ISBS, Proceeding, Lisboa, Portugal, pp. 321–325 (1996)
26. Akça, F.: Prediction of rowing ergometer performance from functional anaerobic power, strength and anthropometric components. J. Hum. Kinet. **41**(1), 133–142 (2014)
27. Kleshnev, V.: Race strategy on the last worlds: race tactics in the finalist and countries specifics. Marker on the oar shaft and blade depth. Rowing Biomech. Newslett. **1**(10) (2001)
28. Secher, N.H.: Physiological and biomechanical aspects of rowing. Sports Med. **15**(1), 24–42 (1993)
29. Kleshnev, V.: Moving the rowers: biomechanical background. Aust. Rowing Carine WA **25**, 16–19 (2002)
30. Sliasas, A., Tullis, S.: Modelling the effect of oar shaft bending during the rowing stroke. Proc. Inst. Mech. Eng. Part P: J. Sports Eng. Technol. **225**(4), 265–270 (2011)
31. Kleshnev, V.: Segments contribution to the stroke length and rowing power. Higher percentage of trunk power b help to achieve higher boat speed. Rowing Biomech. Newslett. **2**(2) (2002)
32. Sliasas, A., Tullis, S.: Numerical modelling of rowing blade hydrodynamics. Sports Eng. **12**(1), 31–40 (2009)
33. Baudouin, A., Hawkins, D.: Investigation of biomechanical factors affecting rowing performance. J. Biomech. **37**(7), 969–976 (2004)
34. Pilgeram, K.C., Delwiche, M.J.: Device for on-the-water measurement of rowing output. Sports Eng. **9**(3), 165–174 (2006)
35. Sliasas, A., Tullis, S.: The dynamic flow behaviour of an oar blade in motion using a hydrodynamics-based shell-velocity-coupled model of a rowing stroke. Proc. Inst. Mech. Eng. Part P: J. Sports Eng. Technol. **224**(1), 9–24 (2010)
36. Schwanitz, P.: Applying biomechanics to improve rowing performance. FISA Coach **2**(3), 1–7 (1991)
37. Leroyer, A., et al.: Influence of free surface, unsteadiness and viscous effects on oar blade hydrodynamic loads. J. Sports Sci. **28**(12), 1287–1298 (2010)

Chapter 2
Biomechanics and Hydrodynamic of the Rowing

2.1 Biomechanics of the Rower

Biomechanics is defined as the mechanics of the structure and the functionality of the biological system [1, 2]. Meanwhile, the biomechanics of rowing can be described as the body's functions and the rower's motion sequence during the stroke. Several biomechanics factors affect rowing performance, such as anthropometry, body weight, stroke rate, and rowing style [3–6]. The rowers must be familiar with the stroke technique because it influences propulsive efficiency and improves performance [3, 5, 7, 8]. Besides that, to become a better rower, the rower needs to understand the body's task related to rowing biomechanics [9], specifically using the leg, arm, and trunk [7, 10]. Kleshnev [11] reported that the average amount of power applied during the stroke was $46.4 \pm 4.5\%$ from the legs, $30.9 \pm 5.2\%$ from the trunk, and $22.7 \pm 5.2\%$ from the arms [12]. Changing the percentage of power from any of these body parts alters the force applied, leading to different results [13].

The rowing technique is the key to performance [14–16]. V. Kleshnev reported that the technique could be classified into four styles based on the two most significant body segments: Leg and trunk, as shown in Fig. 2.1. The four rowing techniques are Rosenberg, Adam, DDR, and Grinko style [17, 18]. The characteristics of the different types depend on the body part's timing and emphasis.

Rosenberg's style applies a large forward declination of the trunk, then explosive leg extension without significant trunk bend, and then at the end of the drive, the trunk stops by a considerably backward position. Adam's style then applies a relatively long leg drive with simultaneously active leg and trunk during the stroke. The DDR style then uses significant forward decline, like the Rosenberg style, but simultaneously activates the leg. Finally, the Grinko style applies consequent timing, emphasising the drive phase's leg extension. Although the technique is classified based on the leg and trunk, arms also play an important role. Based on the power applied, arms contribute 22.7% of the overall power of the one-stroke [12, 18]. Besides that, the arms also

© The Author(s), under exclusive license to Springer Nature Singapore Pte Ltd. 2025
A. A. Mohd Yusof et al., *Rowing Biomechanics and Hydrodynamics*,
Series in BioEngineering, https://doi.org/10.1007/978-981-97-7271-1_2

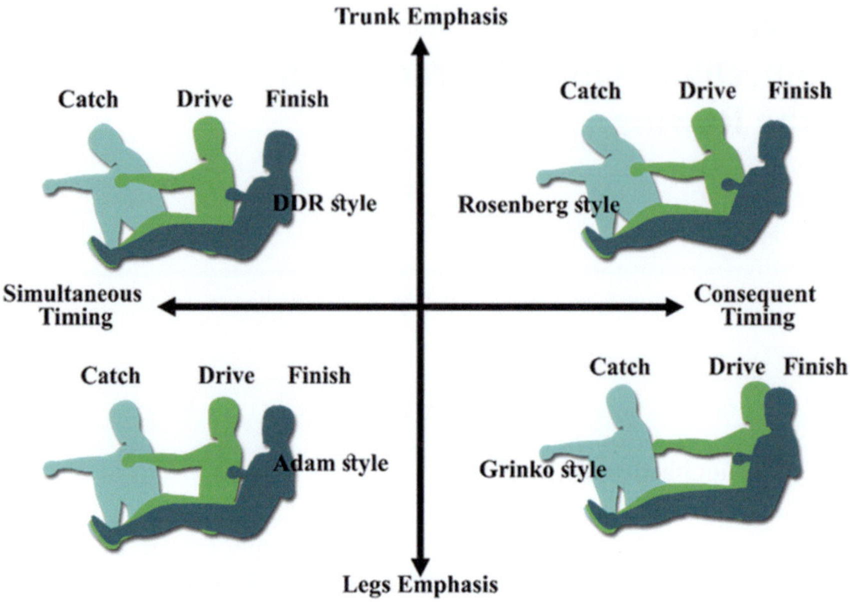

Fig. 2.1 Rowing styles practically applied by the rower adapted from [4]

keep strokes at the right timing for catch and finish, thus avoiding backsplash and early finish [19].

2.2 Rower Force Profile

A musculoskeletal force applied to the oar's handle during a stroke can be visualised as the handle force profile [5, 7, 11, 20]. Smith et al. interpreted the biomechanics of rowers as a force angle graph that provided force information based on the swept angle [21]. A. Baudouin et al. expressed the handle force profile according to the percentage of the stroke to match the graph pattern of the several rowers.

The compilations of handle force profiles from the previous study are illustrated in Fig. 2.2. To make the force profile easier to evaluate, each handle force profile's drive angle or time is changed into a percentage of the drive. The details of the force profile are tabulated in Table 2.1. Generally, the handle force's peak force ranges from 330 to 772 N [5, 8, 21, 22, 22, 23, 23–25]. Smith et al. reported a peak force of 757 N generated [21]. A. Baudouin et al. reported that using a coxless pair boat, 330 N of the peak force was obtained, and the lowest peak was reported compared to the others [8]. K. C. Pilgeram et al. measured the force using a device for on-the-water, and the peak force was 348 N [23]. In their study, M. M. Doyle et al. compared the difference between the heavyweight and lightweight rowers, and the peak force was 507 N [5]. Using an experimental study, v. Kleshnev studied the different measuring forces placed at the oar handle and the pin. It showed that the force measured at both places was similar at 285 N [25]. By referring to the force value of A. Sliasas et al., who adapted the experimental results from V. Kleshnev as input data for the simulation work, reported that the peak profile was 707 N [22]. Finally, V. Kleshnev came out with the peak force profile of 339 N, captured using the bioRowtel device [24].

The peak force achieved was in the range of 30.35–51.99% of the drive. Baudouin et al. [8] and Kleshnev [25] reported that the peak force was reached later than the others, 51.48 and 51.99%. The other papers provided the peak result of the force profiles in 30% and 40% of the drive phase. At 43.57%, 43.57%, and 40.32% of phase drive were reported by Sliasas et al. [22], Pilgeram et al. [23], and Smith et al. [26], respectively. The rowers applied the earliest peak force, 30.35% and 37.59% of the drive phase, as studied by Doyle et al. [5] and Kleshnev [24], respectively.

The peak timing condition is related to the stroke rate. It is found that the average stroke rate applied by the rower is 30 strokes per minute, which ranges from 28 to 36 strokes per minute [6, 27–29]. When lower stroke rates are applied, for instance, 28 strokes per min, the peak force is achieved at 51.48% of the drive phase. However, when the higher stroke rate is applied (30–36 strokes per min), the peak force is achieved earlier (40.32–30.35% of the drive phase). Besides that, Baudouin et al. [8] reported that the stroke rate increased the boat's average velocity.

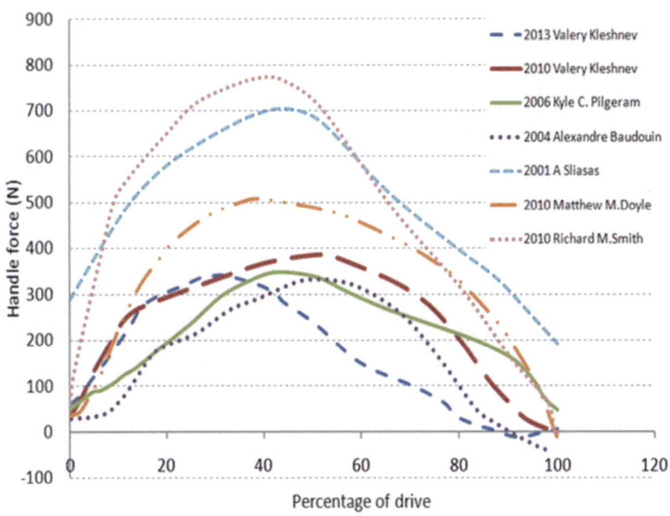

Fig. 2.2 Rower force profile applied during the drive

Table 2.1 The force profiles database applied by the rower during the drive phase

Studies	Measuring method	Max force (N)	Percentage drive to achieve peak	Stroke rate (stroke/min)	Boat type	Total angle rotation (degree)	Blade type
Smith et al. [21]	At rigger using pin force (3D piezoelectrical transducer)	772	40.32%	30	Single scull	106	n/a
Baudouin et al. [8]	Using a strain gauge and potentiometer	330	51.48%	28	Coxless pair	n/a	Concept2 ultralight oars
Pilgeram et al. [23]	Using a strain gauge and potentiometer	348	43.57%	n/a	Boat of four swings and coxswain	84	Big blade concept II
Doyle et al. [5]	Two-dimensional load transducers	507	37.59%	n/a	Men coxless pair	84	n/a
Kleshnev [25]	Using bioRowtel	385	51.99%	n/a	n/a	113	n/a
Sliasas et al. [22]	CFD	707	43.57%	36	Heavy men 4-	n/a	Big blade
Kleshnev [24]	Using bioRowtel	339	30.35%	36	Single scull	n/a	n/a

n/a data not available

2.3 Research Improvement in the Biomechanics of Rower

Kinematic and kinetic differences were identified by Page et al. [30]. The study used the modified rowing ergometer to collect the data. The system improvised the monitoring method and proved its potential as a training tool for the rower and coach. However, two gaps were identified in this study. First, the study does not include the dynamic of a moving boat. Thus, the rower rows at the exact location. Second, the study does not involve the oar; instead, the rower rows by pulling the ergometer handle, producing different rower kinematics and kinetics than on-water training.

Pudlo et al. integrated a 3D motion analysis system with the force sensor to investigate the kinetic and kinematics of the rower [31]. The study was found to be unique and original because the authors analysed the rower motion precisely in three-dimensional axes. However, the applied method simplified the rower biomechanics using an ergometer representing a rowing boat. A similar method was used by Černe et al. [32]. His study aimed to develop and evaluate a system to measure real-time biomechanical analysis of rowers on the ergometer. It was found that there is a noticeable difference in measured parameters between expert and non-expert rowers, which concludes that the applied method can be used to characterise the rowing technique. However, the data was captured using an ergometer, which simplified the rower biomechanics of the rower [32].

The effect of rower kinematics on rowing performance in water studies is hard to capture; thus, most researchers do not include the kinematics aspect except Smith et al. [21]. In the study, the kinematic of the rower was recorded by using the seat's position to estimate the position of the rower's body. The connection between the criterion of success in rowing, the force acting on the boat, and boat speed was demonstrated in the study. In addition, the force profile of the skilled and unskilled rowers was investigated, and differences in force timing were observed. However, the method's effectiveness was low as the seat position could not represent the rower's actual motion, ignoring the kinematic of the trunk and arms. Besides that, minimal information was captured. Pilgeram et al. and Doyle et al. developed a system for the water measurement of the work done during rowing [5, 23]. In Pilgeram et al.'s [23] study, it was found that the heavyweight crews showed significance in examined variables. Meanwhile, Observation data for the lightweight crew is consistent but slower.

Doyle et al. [5] investigated the difference between the heavyweight and lightweight crew and compared the force pattern produced. The study showed that although the power produced by the heavyweight was higher than the lightweight, lower normal force affected the heavyweight rower's power index. Both systems only focused on the handle force measurement, which did not include rower kinematics and the second important aspect of hydrodynamic force, according to Baudouin et al., the pairing correlated and affected total propulsive power, synchronisation, and total drag contribution [8]. However, although it is crucial, the researchers do not touch on kinematic and hydrodynamic aspects.

Kleshnev [25] presented the pattern handle force profile generated by the rower at two different stroke rates [33]. The data showed that higher force was produced at a smaller sweep angle for a higher stroke rate. However, the author did not investigate a hydrodynamic aspect and how the rower moved based on the kinematic element, which was more significant. The details of force and power in rowing were discussed by Kleshnev [20]. The study explained that the amount of impulse and kinetic energy was the fundamental element of the average boat velocity, which affected the competition result.

Laschowsk et al. studied the effect of oar-shaft stiffness and length on rower biomechanics [34]. It found that shaft stiffness was not significant in improving the boat acceleration. The study focused specifically on the oar equipment. However, the effect of the shaft stiffness on the hydrodynamic force of the blade was not considered, which was found to be the research gap.

2.4 Hydrodynamic of the Blades During the Stroke

The hydrodynamics of the blade includes the hydrodynamics force generated by the blade, the peak force reported by the previous research, the blade design, the blade design features, and the force components. Understanding these aspects can enhance the rowing performance by improving blade efficiency and energy conservation, especially during the drive phase.

When the submerged blade is rotated with a specific relative velocity and oriented at a particular projection area, the following equation can define the hydrodynamic force generated on the blade (Eq. 2.4).

$$F_p = \frac{C_p \rho v_r^2 A_o}{2} \tag{2.4}$$

where F_p is a force acting on the blade, C_p is blade coefficients, ρ is water density, v_r is relative velocity, and A_o is the projection area of the blade [10, 35, 36].

Several factors influence the magnitude of the force: (1) surface area, (2) blade coefficient and (3) the relative speed [37–39]. The relative speed is the difference between oar angular and boat translation speeds. The blade coefficient depends on the blade designs, including the features and their position during the stroke. Over the decade, blade designs have been revolutionised in size and shape to improve performance.

The force acting on the blade is obtained from the total pressure generated. The total pressure is defined as follows (Eq. 2.5):

$$P_{total} = \frac{Force}{A_o} = P_{dynamic} + P_{static} \tag{2.5}$$

The total pressure is the sum of static and dynamic pressure, as shown in Fig. 2.3. It is also known as the stagnation pressure, measured at the stagnation point in the fluid. At this point, the velocity of the fluid is zero, and all the kinetic energy is converted into pressure energy through an anisotropic process where no heat is involved. Instead, the transfer of energy happened due to friction [40].

The first component of the total pressure is dynamic pressure [41, 42]. The kinetic energy of the flowing fluid forms the pressure. The total pressure is the function of density, ρ, fluid velocity, v and is inversely proportional to the gravity acceleration, g, as in Eq. 2.6.

$$P_{dynamic} = \rho \frac{v^2}{2g} \tag{2.6}$$

The static pressure refers to the pressure of the still fluid nominated by the depth. The static pressure is measured using the measuring device in which fluid is in static condition or travelling along with the flow [41]. When the blade is submerged, the static pressure acts in all directions as follows;

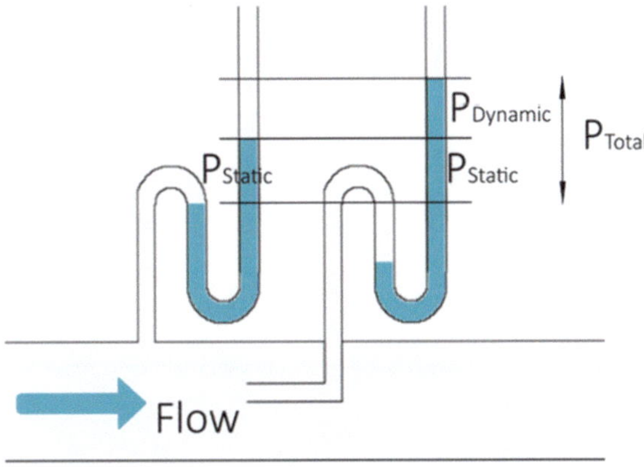

Fig. 2.3 The difference between static, dynamic, and total pressure

$$P_{static} = \rho g h \tag{2.7}$$

where ρ is the fluid density, g is the gravity acceleration, and h is the depth of the fluid.

2.5 Hydrodynamic Force Components

Experimental study of the hydrodynamic performance using in-site measurement is complicated. Many factors may have affected the result, such as yaw, surge, pitch, heave, row, and rotating moments [16, 26, 43]. Thus, most of the investigation is done under controlled conditions. By considering the important aspect of blade hydrodynamics, Sliasas et al. successfully performed computational simulations to evaluate the hydrodynamic force for steady and unsteady flow conditions [37, 44]. The hydrodynamic force was resolved into lift and drag force components. The lift force dominates the propulsive mechanism, contrary to most collected literature that found the drag force overwhelmed the rowing propulsive [39].

The hydrodynamic force of the blade is generated due to the relative speed between the oar angular speed and boat translation speed, as had been studied by Leroyer et al. and Sliasas et al. [39, 45]. Hydrodynamic forces acting on the blade can be resolved according to the oar rotational axis (Fn and Ft) and boat translation axis (Fd and Fl), as illustrated in Fig. 2.4 [39]. An experimental study by Leroyer used a simplified rowing mechanism that shows the pattern of the hydrodynamic force acting on the blade according to the Fx and Fy axis to represent the force direction of the moving boat.

The boat speed increased exponentially in the competition as the oar rotated [45]. However, as the oar's speed passes the peak and gradually decreases as it reaches the finish, the boat's speed maintains due to inertia. The rower applied the force at the oar handle, causing the oar to rotate with a certain angular velocity. The blade acts as

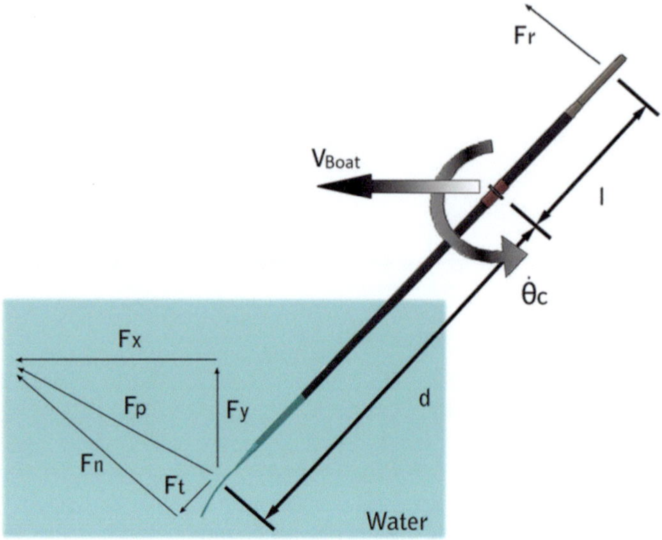

Fig. 2.4 Force components acting on the submerged blade during the drive phase due to the relative speed of oar angular and boat translation speeds

the anchor to transfer the rower energy based on the drag-induce concept mentioned by Kinoshita et al. [46]. However, the energy transformation is less effective due to the blade slip of the drive phase. According to Caplan et al., the problem could be improved by targeting the design of the blade, which can maximise the force generated on the blade [35].

2.6 Calculation of Blade Force

Force acting on the blade is defined as Question 2.4. The force calculation is essential to estimate the magnitude of the force acting on the blade during the drive, reflecting its performance later. For example, to find the best cruise speed and rower capability to row the oar at the best effort for excellent performance. The calculation is based on the information from the previous study by Yusof [47].

It takes the big blade's area of 0.1260 m^2 into account. Based on the study, the angular velocity of the oar blade is 0.88 rad/s, outboard length is 2.66 m, and Vb is 0.495 m/s, where the velocity value during the start of the boat is higher propulsive force generated. The propulsive force coefficient is 1.33 for a water density of 1000 kg m^{-3}. It assumes the rower rows the boat smoothly without such yaw, surge, pitch, heave, row and rotating moments present during the drive. Thus:

$$\text{Oar speed, Voar} = r\omega = 0.88 \times 2.66 = 2.34\,\text{m s}^{-1}$$

The velocity of the oar due to the boat's translation motion and rotational of the oar blade can be solved using the relative velocity of a vector. This calculation considers 60° of oar drive angle from the boat's moving direction.

$$\overleftarrow{V_b} = 0.49i$$
$$\overleftarrow{V_{\text{oar}}} = -2.02i - 1.17\text{j}$$
$$\overleftarrow{V_r} = \overleftarrow{V_{\text{oar}}} + \overleftarrow{V_b} = -1.53i - 1.17j$$

Sample calculations for propulsive force and its components are as follows (Eq. 2.4):

$$Propulsive\,force,\ F_p = \frac{C_p \rho v_r^2 A_o}{2}$$
$$= \frac{1.33 \times 1000 \times 1.92^2 \times 0.126}{2} = 309\,\text{N}$$

The propulsive force can be resolved into the force component depending on the blade position, which is as mentioned in Sect. 1.6.2; the force can be resolved either based on the boat moving direction (x and y-axis) or based on the oar blade moving direction (Fn and Ft).

Take the example of an oar drive angle of 60° from a boat moving direction for F_x and F_y, as shown in Fig. 2.5.

Due to the relative speed of the boat speed and oar, the direction of the propulsive force directs a bit outward at 37° based on the vector operation. Using this angle

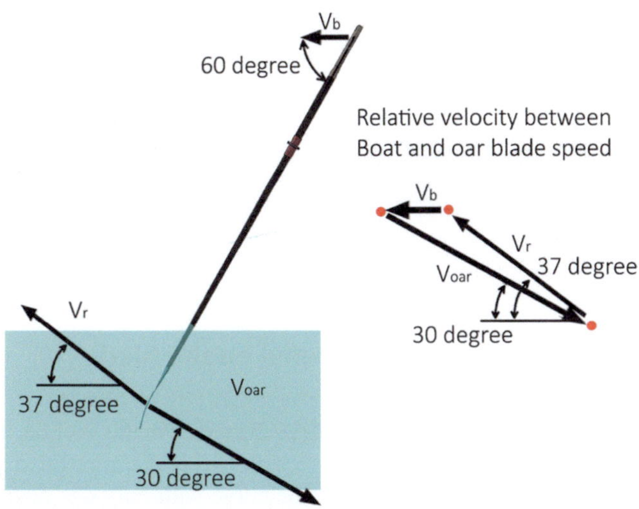

Fig. 2.5 Force components acting on the submerged blade during the drive phase at 60°

value, the magnitude of the forces is:

$$\begin{aligned} F_x &= F_p x \cos \theta_{\text{oar}} \\ &= 309\,\text{N} x \cos 37 = 247\,\text{N} \end{aligned} \tag{2.8}$$

$$\begin{aligned} F_y &= F_p x \sin \theta_{\text{oar}} \\ &= 309 N x \sin 37 = 186\,\text{N} \end{aligned} \tag{2.9}$$

Deflection of the angle from the oar axis

$$\begin{aligned} \Delta\theta_{\text{oar}} &= \theta_{\text{oar}} - \dot{\theta}_{\text{oar}} \\ &= 37 - 30 = 7\,\text{degree} \end{aligned} \tag{2.10}$$

Thus, force components in the oar blade axis, Fn and Ft, are as follows:

$$\begin{aligned} F_n &= 309 N x \cos 7 = 307\,\text{N} \\ F_t &= 309 N x \sin 7 = 37\,\text{N} \end{aligned}$$

In other words, Fn can be represented as drag force, Fd acts normally to the oar blade, whereas Ft can be described as lift force, Fl acts parallel to the oar blade. Using the modification equation as a propulsive force, drag and lift coefficient at 60 degrees can be obtained.

$$\begin{aligned} C_d &= \frac{2F_d}{\rho v_r^2 A_o} \\ &\frac{2 \times 307}{1000 \times 1.92^2 \times 0.126} = 1.32 \end{aligned} \tag{2.11}$$

$$\begin{aligned} C_l &= \frac{2F_l}{\rho v_r^2 A_o} \\ &\frac{2 \times 37}{1000 \times 1.92^2 \times 0.126} = 0.159 \end{aligned} \tag{2.12}$$

2.7 Lift and Drag Force Generated on the Blade

Lift and drag forces are essential aspects to be discussed in this section. The lift force is illustrated as a sinusoid, repetitive curve with two peaks in positive and negative forms. In contrast, the drag force is a cosine curve with only one positive peak. In this scope of the study, the lift force is defined as the force acting perpendicular to the direction of the blade relative velocity. The drag is then defined as the force acting parallel to the direction of the blade's relative velocity.

Using CFD simulation, Sliasas et al. reported that the maximum hydrodynamic forces generated were 393 and 491 N [37, 44]. Leroyer et al. experimented using a fabricated rowing device to study the hydrodynamics of rowing. The computational simulation was then performed according to the experimental test [39]. The results showed that the maximum lift was 51.66 and – 48.20 N. Drag forces were 104.44 N. The CFD result also showed a pattern similar to that of the experiments, with the maximum lift force of 58.60 and – 45.90 N. The maximum drag force of 97.68 N was reported. Barre et al. then used the same device to compare forces on the blade with a towing tank [48]. The result was presented as a hydrodynamic force for the 0.0420 m^2 projection areas of the flat blade and 0.0361 m^2 curved blades; the generated forces were 58 N and 52 N, respectively. Table 2.2 summarises all the explained data.

2.8 Force Coefficient of the Blade

The force coefficient is a dimensionless parameter used to qualify the characteristic of a particular shape [35]. It is the force-generated ratio to the fluid density, velocity, and blade projection area. Most papers resolve the blade coefficient into the lift and drag coefficient, which is classified according to the force-acting direction (refer to Table 2.3). Five types of blades in the previous papers work on the force coefficient: flat rectangular, curved rectangular, Macon, flat Hatchet, and curved Hatchet [35, 36, 45, 49, 50]. The Hatchet blade, also known as the Big blade among the rowing community, is manufactured by Concept2 Company. However, to make it consistent with the whole paper, the name Big Blade was selected for the research explanation. The coefficient depends on the density of the water, the relative velocity between the blade and fluid, and the projection area of the blade. According to the five journal papers reviewed in this area, the coefficient ranges from 2.53 to 1.33 for the lift and 152 to – 0.45 for the drag. The coefficient includes all blade scales, quarter and full scale, whose detail is given in Table 2.3 [35, 36, 45, 49, 50].

2.9 Blade Design and Features

The blade is the oar component used to generate hydrodynamic force. There are three types of commercial blade designs currently used by the rower: (1) Macon Blade, (2) Big blade, and (3) Fat blade [35, 51, 52]. Each blade has its own identified features that influence its performance. The Macon blade is the oldest design, introduced in 1959. After 32 years, it is Big blade's turn to be introduced. The big blade model proves this new design could improve the rowing performance compared to the previous blade. The latest version of the oar blade was introduced in 2004, known as the Fat blade [52]. The Macon blade is designed to be symmetrical, where the shape

Table 2.2 Database of hydrodynamic force generated on the blade

Studies	Measuring method	Total sweep angle	Blade type	Projection area (m^2)	Propulsive force (N)	Max. lift force (N)	Max. drag force (N)
Leroyeret al. [39]	Fabricated rowing device/CFD	180	Flat rectangular	0.0400	n/a	51.66 and − 48.20 (experimental)	104.44 (experimental)
					n/a	58.60 and − 45.9 (simulation)	97.68 (simulation)
Sliasas et al. [44]	CFD	360	Flat big blade	0.1260	393	309.00	314.00
Barré et al. [49]	Fabricated rowing device	360	Flat	0.0420	58	n/a	n/a
			Spherical	0.0361	52	n/a	n/a
Sliasas et al. [37]	CFD	360	Flat big blade	0.1260	491	400.00	273.00

n/a data not available

Table 2.3 Database of the force coefficient related to the blade hydrodynamic

Studies	Blade moving profile	Measuring method	Blade shape	Projection area (m^2)	Drag coefficient	Lift coefficient	
Caplan et al. [35]	Quasi-static	Strain gauge	Flat rectangular	0.00773	1.96	1.36	– 1.21
			Curved rectangular	0.00775	2.06	1.52	– 1.43
			Flat big blade	0.00774	1.932	1.28	– 1.10
			Curve big blade	0.00774	2.15	1.29	– 1.31
Sliasas et al. [45]	Quasi-static	CFD	Curved rectangular	0.00785	1.95	1.22	– 1.25
			Curved rectangular	0.12600	1.33	0.94	– 0.96
Coppel et al. [36]	Quasi-static	CFD	Flat plate	0.007742	2.53	1.47	– 1.45
			Curved big blade	0.007741	1.86	1.85	– 1.32
			Curved big blade	0.12400	1.38	1.10	– 1.61
Barré et al. [49]	Transient and rotate	The fabricated device with servo and dynamometer	Flat	0.04200	1.98	0.87	– 0.91
			Spherical	0.03610	2.15	0.97	– 0.94
Coppel et al. [50]	Quasi-static	CFD	Macon	0.00670	1.97	0.94	– 0.47
			Macon	0.10800	1.66	1.02	– 0.45

is elongated. The shaft is then attached to the middle of the blade. The big blade is designed asymmetrically, where the blade's tip is wider than the Macon blade. The fat blade is also designed asymmetrically, where the shaft is attached at the top corner of the blade.

The middle part of the blade is intended to be wider than the tip. However, the performance of blade designs is not fully explored, especially for the latest blade design. The previous study by Caplan et al. reported the investigation of the fluid dynamic characteristic for both Macon and Big blade design, which leads to the finding that the Big blade improves hydrodynamic efficiency under quasi-static conditions by about 2% compared to Macon blade [53]. However, no studies have yet been reported on the performance of Fat blades. Furthermore, the hydrodynamic performance of the blade is under dynamic conditions.

According to Caplan et al., the blade design is not optimised, and there is a gap for improvement [35]. A comparative study of the lift and drag coefficient values between rectangular and Big-blade models showed differences. The rectangular blade model generates higher force [35, 36]. Macon blade then developed low propulsive force and higher hydro resistance. It is designed with a lower aspect ratio and symmetry [35, 50, 54].

The fluid flow around the blade was studied by Leroyer et al. [39]. The blade motion was reproduced by mimicking the motion of the angular speed of the oar and boat translation speed. The force's magnitude depends on the oar's relative velocity and boat speed. A mutual agreement was found between computational and experimental results. Throughout the carried study, the water flow around the blade was complex and involved violent free-surface including elevation of wave and cavity at the back of the blade [38, 39]. Sliasas et al. proved the effect of the turbulence during the drive phase by vortex formation near to blade tip for strengthening the pressure generated [37]. Meanwhile, blade design features improve rowing propulsion by changing the fluid flow [35].

Blade performance is enhanced throughout the introduction of the shape design features. It is found that the curved blade performs better because more water gets trapped on its face [35]. Performance improvement happens due to the alteration of the fluid flow aspect, which is influenced by turbulence formation [36]. The lift and drag coefficient is inversely proportional to the size and area of the blade, in which the coefficient value reduces as the blade area increases. Caplan et al. reported that the curve parameter increases the lift force at the rotation angle from 0° to 15°, 40° to 70°, and 90° to 180°. The curve parameter could also improve the blade slip problem of the drive phase, leading to increased blade efficiency [35].

2.10 Researcher Improvement in Hydrodynamic of the Blade

The study on blade hydrodynamics usually focused on the pressure generated during the stroke, fluid flow around the blade, blade shape features, and related force coefficient. Caplan et al. examined the influence of blade design on rowing performance using a quasi-static experimental study [35]. The study accounts for the complex interaction between blade and water and presents the force coefficient result. However, the simplified propulsive mechanism applied to the methodology ignored blade dynamic and rower biomechanics factors.

Using the mathematical model, Caplan et al. compared blade designs by considering the relative movement between the blade and water to estimate the boat velocity [55]. In the same year, 2007, Caplan et al. investigated the fluid characteristics of Macon and Big Blade's commonly used blade model [53]. It was found that the Big blade exceeded the Macon blade by 2% of the blade's hydrodynamic performance. The study only focused on the blade shape, which scales down the related parameters involved in the study. The simplification working mechanism of the full-scale blade is not comprehensive.

Leroyer et al. used CFD to verify the Reynolds-Average Navier–Stokes equation's capabilities to compute the blade's fluid flow [39]. This approach used the dynamic motion of the blade, which the blade rotated and translated at once. However, the motions did not represent the actual blade motion condition in which it was rotated around 0–180° at a constant translation speed. In addition, this study ignored human factors and used smaller and simplified blade shapes. Sliasas et al. used blade hydrodynamic data from Caplan et al. to proceed with the study using CFD [35, 37]. The blade was set to rotate according to competition for the actual blade size. From the result of the blade-water interaction, the flow was divided into six regimes depending on the relative blade motion and the influence of the lift and drag generated. The study used an almost similar setting to Leroyer et al.; which blade was rotated without considering the blade catch angle is one of the primary data of rower stroke [39].

Barre et al. described the hydrodynamic force of the blade and compared it with the standard model using the experimental method [49]. The study used a rowing device with 0.7 scales on the blade. It highlighted that the validation of the study towards the actual blade kinematic was questionable, as it did not conclude the variation of the boat speed during the stroke. Leroyer et al. published the second journal. Continuing the previous study on the influence of free surface and unsteadiness and viscous effects on the flow around the blade, free surface must be considered in CFD, where the viscosity effect had minimal influence on the result [56]. The study used experimental and simulation approaches with simplification motion, which did not consider dynamic blade conditions due to the rower-powered stroke.

The comparisons between the blade models were made by Coppel et al. using CFD analysis [36]. The investigation provided the preliminary result to the fluid simulation around the full-scale blade by considering dynamic similarity matching between the quarter-scale and full-scale blades. The paper proved that dynamic similarity between the quarter and full-scale blade could be achieved using two flow velocities of 20 m/s and 5 m/s, respectively. The study improvised the previous research by considering the full blade size. However, boundary conditions and blade dynamic motion were not included.

References

1. Alexander, R.M.: Mechanics of animal movement. Curr. Biol. **15**(16), R616–R619 (2005)
2. Hatze, H.: The meaning of the term 'biomechanics.' J. Biomech. **7**(2), 189–190 (1974)
3. Barrett, R.S., Manning, J.M.: Relationships between rigging set-up, anthropometry, physical capacity, rowing kinematics and rowing performance. Sports Biomech./Int. Soc. Biomech. Sports **3**(2), 221–235 (2004)
4. Kleshnev, V.: Rowing styles. Rowing Biomech. Newslett. **6**(60) (2006)
5. Doyle, M.M., Lyttle, A., Elliott, B.: Comparison of force-related performance indicators between heavyweight and lightweight rowers. Sports Biomech. **9**(3), 178–192 (2010)
6. Kleshnev, V.: Stroke rate vs. distance in rowing during the Sydney Olympics. Aust. Inst. Sport Aust. Rowing **24**(2), 18–22 (2001)
7. Baudouin, A., Hawkins, D.: A biomechanical review of factors affecting rowing performance. Br. J. Sports Med. **36**(6), 396–402 (2002)

8. Baudouin, A., Hawkins, D.: Investigation of biomechanical factors affecting rowing performance. J. Biomech. **37**(7), 969–976 (2004)
9. Nolte, V.: Introduction to the biomechanics of rowing. FISA Coach **2**(1), 1–6 (1991)
10. Caplan, N., Coppel, A., Gardner, T.: A review of propulsive mechanisms in rowing. Proc. Inst. Mech. Eng. Part P: J. Sports Eng. Technol. **224**(1), 1–8 (2010)
11. Kleshnev, V.: Vertical seat force; lift force and the height of the handle and foot-stretcher. Optimal height of the handle; push foot-stretcher with toes at catch; horizontal drive. Rowing Biomech. Newslett. **2**(5) (2002)
12. Kleshnev, V.: Segments contribution to the stroke length and rower power. Higher percentage of the trunk power helps to achieve higher boat speed. Rowing Biomech. Newslett. **2**(2) (2002)
13. Kleshnev, V.: Experimental data on comparison of rowing on-water with Concept-II and RowPerfect ergometers. Rowing Biomech. Newslett. **5**(3) (2005)
14. Soper, C., Hume, P.A.: Towards an ideal rowing technique for performance. Sports Med. **34**(12), 825–848 (2004)
15. Sanderson, B., Martindale, W.: Towards optimizing rowing technique. Med. Sci. Sports Exerc. **18**(4), 454–468 (1986)
16. Formaggia, L., et al.: A three-dimensional model for the dynamics and hydrodynamics of rowing boats. Proc. Inst. Mech. Eng. Part P: J. Sports Eng. Technol. **224**(1), 51–61 (2010)
17. Kleshnev, V.: Power in rowing. International Research in Sports Biomechanics, pp. 224–230 (2002)
18. Kleshnev, V.: Biomechanics of rowing. Rowing Faster, 2nd edn. Serious training for serious rowers. Human Kinetics, pp. 105–121 (2011)
19. Macrossan, M.N., Macrossan, N.W.: Back-splash in rowing-shell propulsion (2006)
20. Kleshnev, V.: Force and power in rowing. Rowing Biomech. Newslett. **14**(156) (2014)
21. Smith, R.M., Loschner, C.: Biomechanics feedback for rowing. J. Sports Sci. **20**(10), 783–791 (2002)
22. Sliasas, A., Tullis, S.: Modelling the effect of oar shaft bending during the rowing stroke. Proc. Inst. Mech. Eng. Part P: J. Sports Eng. Technol. **225**(4), 265–270 (2011)
23. Pilgeram, K.C., Delwiche, M.J.: Device for on-the-water measurement of rowing output. Sports Eng. **9**(3), 165–174 (2006)
24. Kleshnev, V.: Vertical handle force. Rowing Biomech. Newslett. **13**(143) (2013)
25. Kleshnev, V.: Comparison of measurements of the force at the oar handle and at the gate or pin. Rowing Biomech. Newslett. **10**(108) (2010)
26. Smith, R., Draper, C.: Quantitative characteristics of coxless pair-oar rowing. In: ISBS-Conference Proceedings Archive (2002)
27. Kleshnev, V.: Propulsive efficiency of rowing. In: Proceedings of the XVII International Symposium on Biomechanics in Sports (1999)
28. Kleshnev, V.: The effect of stroke rate on biomechanical parameters and efficiency of rowing. In: XIV Symposium ISBS, Proceeding, Lisboa, Portugal, pp. 321–325 (1996)
29. Schwanitz, P.: Applying biomechanics to improve rowing performance. FISA Coach **2**(3), 1–7 (1991)
30. Page, P.N., Hawkins, D.: A real-time biomechanical feedback system for training rowers. Sports Eng. **6**(2), 67–79 (2003)
31. Pudlo, P., Pinti, A., Lepoutre, F.: Experimental laboratory apparatus to analyze kinematics and 3D kinetics in rowing. Sports Eng. **8**(1), 39–46 (2005)
32. Černe, T., Kamnik, R., Munih, M.: The measurement setup for real-time biomechanical analysis of rowing on an ergometer. Measurement **44**(10), 1819–1827 (2011)
33. Coppel, A.L.: A computational fluid dynamic investigation of rowing oar blades. PHD Thesis, The University of Birminghan, School of Sport and Exercise Sciences (2010)
34. Laschowski, B., et al.: The effects of oar-shaft stiffness and length on rowing biomechanics. Proc. Inst. Mech. Eng. Part P: J. Sports Eng. Technol. 1754337115582121 (2015)
35. Caplan, N., Gardner, T.N.: Optimization of oar blade design for improved performance in rowing. J. Sports Sci. **25**(13), 1471–1478 (2007)

36. Coppel, A., et al.: Simulating the fluid dynamic behaviour of oar blades in competition rowing. Proc. Inst. Mech. Eng. Part P: J. Sports Eng. Technol. **224**(1), 25–35 (2010)
37. Sliasas, A., Tullis, S.: The dynamic flow behaviour of an oar blade in motion using a hydrodynamics-based shell-velocity-coupled model of a rowing stroke. Proc. Inst. Mech. Eng. Part P: J. Sports Eng. Technol. **224**(1), 9–24 (2010)
38. Macrossan, M.N., Kamphorst, M.: Computational study of the Froude number effects on the flow around a rowing blade (2009)
39. Leroyer, A., et al.: Experimental and numerical investigations of the flow around an oar blade. J. Mar. Sci. Technol. **13**(1), 1–15 (2008)
40. Pati, S., A Textbook on Fluid Mechanics and Hydraulic Machines. Tata McGraw Hill (2012)
41. Benenson, W., et al.: Handbook of Physics. Springer (2006)
42. Johnson, R.W.: Handbook of Fluid Dynamics. CRC Press (2016)
43. Smith, R., Draper, C.: Controlling yaw in a pair rowing boat. In: ISBS-Conference Proceedings Archive (2008)
44. Sliasas, A., Tullis, S.: A hydrodynamics-based model of a rowing stroke simulating effects of drag and lift on oar blade efficiency for various cant angles. Procedia Eng. **2**(2), 2857–2862 (2010)
45. Sliasas, A., Tullis, S.: Numerical modelling of rowing blade hydrodynamics. Sports Eng. **12**(1), 31–40 (2009)
46. Kinoshita, T., et al.: Rowing velocity prediction program with estimating hydrodynamic load acting on an oar blade. In: Kato, N., Kamimura, S. (eds.) Bio-mechanisms of Swimming and Flying, pp. 345–359. Springer, Japan (2008)
47. Yusof, A.A.M., et al.: Analysis of oar blade hydrodynamics for rowing propulsive mechanism: experiment and computational fluid dynamics simulation. Int. Rev. Mech. Eng. (IREME) **12**(11), 920–927 (2018)
48. Barre, S., Kobus, J.-M.: Comparison between common models of forces on oar blades and forces measured by towing tank tests. Proc. Inst. Mech. Eng. Part P: J. Sports Eng. Technol. **224**, 37–50 (2010)
49. Barré, S., Kobus, J.: Comparison between common models of forces on oar blades and forces measured by towing tank tests. Proc. Inst. Mech. Eng. Part P: J. Sports Eng. Technol. **224**(1), 37–50 (2010)
50. Coppel, A., et al.: Numerical modelling of the flow around rowing oar blades (P71). In: The Engineering of Sport, vol. 7, pp. 353–361. Springer (2008)
51. Coppel, A., Gardner, T., Caplan, N., Hargreaves, D.: Oar blade force coefficients and a mathematical model rowing. In: 27 International Conference on Biomechanics in Sports (2009)
52. Nolte, V.: Shorter oars are more effective. J. Appl. Biomech. **25**(1) (2009)
53. Caplan, N., Gardner, T.N.: A fluid dynamic investigation of the Big Blade and Macon oar blade designs in rowing propulsion. J. Sports Sci. **25**(6), 643–650 (2007)
54. Abbott, I.H.: Theory of Wing Sections, Including a Summary of Airfoil Data. Courier Dover Publications (1959)
55. Caplan, N., Gardner, T.: A mathematical model of the oar blade–water interaction in rowing. J. Sports Sci. **25**(9), 1025–1034 (2007)
56. Leroyer, A., et al.: Influence of free surface, unsteadiness and viscous effects on oar blade hydrodynamic loads. J. Sports Sci. **28**(12), 1287–1298 (2010)

Chapter 3
The Coupling Mechanism of Rowing

3.1 The Importance of Coupling Between Biomechanics and Hydrodynamics of Rowing

Several studies reported force profile, biomechanics influenced factor, rower kinematic, flow around the blade, and blade design towards the performance [1–6]. However, the studies of those aspects are done separately, although the real rowing system combines all of them. Thus, there is missing information that the coupling mechanism between the rower and oar blade needs to be considered, especially in the study. Therefore, a new comprehensive analysis is necessary to deepen our knowledge about rowing.

The existing blade designs used in the competition indicate performance improvement, especially for Macon and Big Blades. However, no paper has been reported on the fat blade. A previous study states that under the quasi-static condition, the Big blade is assessed to improve performance by about 2% higher than the Macon blade [15]. Unfortunately, only a few studies compare the performance of each blade design specifically under the dynamic condition in which hydrodynamic force is generated due to the relative speed between the oar and boat. Therefore, the computational simulation applied in the study improves the understanding of the blade propulsive mechanism. The method provides insight into the blade's hydrodynamic force, which moves dynamically due to the rower stroke of the phase of the drive.

The rowing style can influence the rowing performance by optimising the rower biomechanics [1, 7]. The study explores the differences in rowing styles. It contributes to deepening the knowledge to enhance the performance by helping the rower maximise the power produced and minimise the energy loss [8, 9].

Finally, the development of methods and analysis brings rowing significant progress in biomechanics and hydrodynamics. The coach does not need to rely solely on his or her eyes to evaluate rower performance, as has been applied in the training session [10]. Besides that, introducing the proposed study would help bring this

© The Author(s), under exclusive license to Springer Nature Singapore Pte Ltd. 2025

A. A. Mohd Yusof et al., *Rowing Biomechanics and Hydrodynamics*,
Series in BioEngineering, https://doi.org/10.1007/978-981-97-7271-1_3

sports technology to new levels in sports engineering. Further research may transpire a definition of the ideal force profile and other parameters. The efforts of the rowing crew could be reviewed after the assessment session. The method could also provide the rower and coach with important technical information. Besides, monitoring stroke timing and force generated. Optimisation of force application could also be analysed, and the rower could inspect his or her performance for each stroke.

3.2 Rowing Oar Mechanism

Oar is the equipment used to propel the boat. On average, the length of the sweep oar is 374 cm, and the length of the sculling oar is 290 cm [11]. It consists of a handle, shaft, button, sleeve, and blade, as shown in Fig. 3.1. The oar shaft is divided into two sections, inboard and outboard, referring to the position of the boat. It is attached to the boat via the oarlock, which allows the oar to rotate freely during the stroke.

The working mechanics of the oar are not fully understood. Some researchers said that oar works as a first-class lever [12]. According to the International Rowing Federation (FISA), oar works as a second-class lever [13]. These two different options were concluded because they view oar working mechanics differently. The first-class lever was reported where force is applied at the handle, and the oar is rotating, revolving the oarlock and hitting the water as a load defined based on the rower perspective.

Meanwhile, the second-class lever is expressed as researchers view rowing as one whole system. The researchers see the blade as the fulcrum that acts as an anchor to drive the boat forward. The best strategy to identify the working mechanism of the oar is based on the force magnitude. Then, the conclusion of the lever class could be made according to the obtained data on mechanical advances.

The working principle of the oar as the lever is illustrated in Fig. 3.2. There are three main elements related to the lever: the fulcrum, load, and effort [14]. Based on the previous explanation, two types of lever classes are recognised. The first type is the first-class lever where fulcrum, FO acts at the middle of the oar, effort acts at the oar handle, F_r and load, F_p acts at the blade [15, 16]. The working principle of this lever causes the oar free to rotate by pivoting at the oarlock at an angular speed of

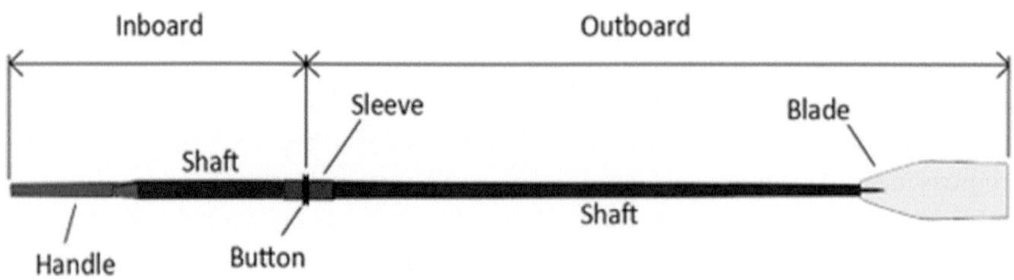

Fig. 3.1 Components and parts of rowing oar

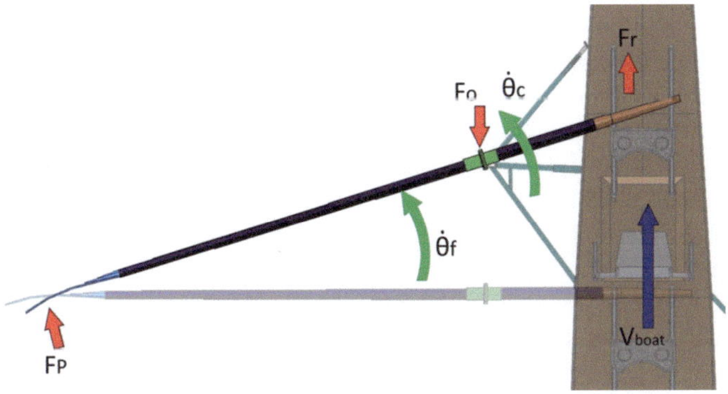

Fig. 3.2 Oar installations on the rigging, force applied, oar motion, and force acting

$\dot{\theta}_c$. The second type is the second class lever, where the fulcrum acts at the blade, F_P acts as the imaginary fulcrum, effort, Fr acts at the handle and load, F_O at the middle of the oar to accelerate the whole rowing boat, and rower by the oar rotated at the imaginary fulcrum with a certain angular angle of $\dot{\theta}_f$. The working principle of this lever causes the whole rowing system to move forward [15, 16].

The oar connects biomechanics and hydrodynamic Rowing [17–20]. The rower applies the force to the oar handle during the drive, producing a hydrodynamic force on the submerged blade. The rower-oar-boat system can be described as a coupling mechanism between the rower, oar, and boat that has the mechanical linkage to transmit the biomechanical force of the rower into hydrodynamic force via the oar [21, 22]. The oar's design features, rigging gearing, and shaft properties affect the oar's efficiency [1, 18, 19]. Previously, D. Cabrera et al. defined oar efficiency according to propulsive efficiency, the ratio of boat dissipation to the rower works [23]. Pendergast et al. described efficiency according to the Froude efficiency. The ratio of total power gives the Froude efficiency dissipated against the drag to the external mechanical power produced by the rower [24].

Meanwhile, Atkinson defined efficiency based on the work done to move the boat to work applied by the rower [25]. Most of the studies that examined oar efficiency have directly related the work done by the rower to the boat. However, there is yet a specific point to the work done at the blade hydrodynamic force.

3.3 The Coupling Mechanism of Rowing

Previous researchers have explored the rower's root forces profile, physiological and anthropometry, fluid dynamic around the blade, and blade design optimisation [1, 3, 26–30]. However, the coupling mechanism between the rower's biomechanics and the blade's hydrodynamic during the stroke received less attention. Biomechanics is interested in figuring out how the rower converts his physiological capacity to propel the boat using hydrodynamic force. Most of these studies that focus on the blade have simplified the rowing stroke mechanism, which causes the analysis to be regarded as not optimum. The deficiency is due to the rowing system's complexity, which leads to the undefined study of biomechanics and hydrodynamics. Moreover, most researchers tend to analyse the biomechanics aspect directly with the boat speed by neglecting the hydrodynamics effect on the oar blade or the other way around.

Limited research on the fluid flow around the oar blade considers the factor of human power [2, 27, 32–34]. A study of this topic would elucidate the unsteady fluid mechanics around the oar blade of the drive phases. In the competition, there are three types of commonly used blades. However, the issue has been raised, and the question that needs to be answered is how the blade works. It is the lift-induced or drug-induced mechanism and how the force is generated [21, 27, 35, 36]. This issue should be questioned since several studies focus on the blade, but the conclusions differ. Besides that, this study also allows further investigation into areas for optimisation and improvement. The blade designs have been changed several times. However, its performance has not been tested for a detailed qualitative assessment of what would constitute a practical design and utilisation.

Besides that, the biomechanics evaluation of the rower stroke is limited, and the related studies do not cover the effect of the stroke styles on rowing performance. Usually, on-the-water evaluation, a method known as 'seat racing', is used [4, 10, 37]. It is done by using a boat of four, where the boats are lined up to race against each other. After the established and fixed distance, the coach would measure the disparity between the two boats based on the racing time. The technique assessment is not precise because the evaluation depends on the coach's experience, and judgment is based on visual consideration. Previously V. Kleshnev reported that the four rowing styles are classified according to body position, timing, and segment emphasis [7]. Classification is an excellent approach to enhancing rowing biomechanics. However, the advantages of each style in terms of performance are not reported, which leaves a big question.

3.4 Assumptions in Investigating Rowing Biomechanics and Hydrodynamic

A dynamic rowing simulator replicates the actual rowing mechanics under control conditions in investigating the rowing's biomechanics and hydrodynamics. The study is carried out using two main methods: experimental and computational. The experimental study obtains the rower handle and blade force, kinematics, angular speed of the oar, and the boat translation speed. The oar angular and boat translation speeds are then assigned as input for the computational study. The scope of the study is simplified as the following list:

1. Boat motion is controlled using a rail, and the boat is allowed to move with one degree of freedom.

2. Biomechanics of Rowing focuses on the single rower for the sweep type oar and boat.
3. Rower's kinematics are captured using video and analysed using motion software.
4. The biomechanics of the rower are fixed to the handle force profile and rower kinematics for each rowing style.
5. Blade designs and features focus only on Macon, Big, and Fat commercial blade models.
6. Computational fluid dynamics (CFD) and experimental studies assess the effect of rowing styles on the blade's hydrodynamics.

References

1. Barrett, R.S., Manning, J.M.: Relationships between rigging set-up, anthropometry, physical capacity, rowing kinematics and rowing performance. Sports Biomech. Int. Soc. Biomech. Sports **3**(2), 221–235 (2004)
2. Caplan, N., Gardner, T.N.: A fluid dynamic investigation of the Big Blade and Macon oar blade designs in rowing propulsion. J. Sports Sci. **25**(6), 643–650 (2007)
3. Baudouin, A., Hawkins, D.: Investigation of biomechanical factors affecting rowing performance. J. Biomech.Biomech. **37**(7), 969–976 (2004)
4. Pilgeram, K.C., Delwiche, M.J.: Device for on-the-water measurement of rowing output. Sports Eng. **9**(3), 165–174 (2006)
5. Leroyer, A., et al.: Influence of free surface, unsteadiness and viscous effects on oar blade hydrodynamic loads. J. Sports Sci. **28**(12), 1287–1298 (2010)
6. Page, P.N., Hawkins, D.: A real-time biomechanical feedback system for training rowers. Sports Eng. **6**(2), 67–79 (2003)
7. Kleshnev, V.: Rowing styles. Rowing Biomech. Newslett. **6**(60) (2006)
8. Kleshnev, V.: Racing strategy in rowing during Sydney Olympic Games. Aust. Rowing **24**(1), 20–23 (2001)
9. Kleshnev, V.: Trends of the race strategy over the last 10 years in different placeholders; prognosis till 2008. Optimal race strategy. Rowing Biomech. Newslett. **2**(19) (2002)
10. Koutedakis, Y.: The role of physiological assessment in team selection with special reference to rowing. Br. J. Sports Med. **23**(1), 51 (1989)
11. Oar Length. http://www.concept2.com/oars/oar-options/length
12. Kleshnev, V.: Biomechanics of rowing. Rowing Faster. Second edition. Serious training for serious rowers. Human Kinetics, pp. 105–121 (2011)
13. Sport, R., Paintworks, D.: FISA Rule Book
14. Rao, C.L., et al.: Engineering Mechanics: Statics and Dynamics. PHI Learning Pvt. Ltd. (2003)
15. Sclater, N., Chironis, N.P.: Mechanisms and Mechanical Devices Sourcebook, vol. 3. McGraw-Hill, New York (2001)
16. McLester, J., Pierre, P.S.: Applied Biomechanics: Concepts and Connections. Nelson Education (2007)
17. Coppel, A., Gardner T., Caplan, N., Hargreaves, D.: Oar blade force coefficients and a mathematical model rowing. In: 27 International Conference on Biomechanics in Sports (2009)
18. Sliasas, A., Tullis, S.: Modelling the effect of oar shaft bending during the rowing stroke. Proc. Inst. Mech. Eng. Part P: J. Sports Eng. Technol. **225**(4), 265–270 (2011)
19. Nolte, V.: Shorter oars are more effective. J. Appl. Biomech. **25**(1) (2009)
20. Fuss, F.K., et al.: Smart oar blade for hydrodynamic analysis of rowing. Procedia Eng. **147**, 735–740 (2016)

21. Kinoshita, T., et al.: Rowing velocity prediction program with estimating hydrodynamic load acting on an oar blade. In: Kato, N., Kamimura, S. (eds.) Bio-mechanisms of Swimming and Flying, pp. 345–359. Springer, Japan (2008)
22. Cabrera, D., Ruina, A., Kleshnev, V.: A simple 1+ dimensional model of rowing mimics observed forces and motions. Hum. Mov. Sci. **25**(2), 192–220 (2006)
23. Cabrera, D., Ruina, A.: Propulsive efficiency of rowing oars. J. Appl. Biomech. (2006, submitted)
24. Pendergast, D., et al.: Energy balance of human locomotion in water. Eur. J. Appl. Physiol. **90**(3), 377–386 (2003)
25. Atkinson, W.C.: Oar blade efficiency. Retrieved Feb. 1, 2005 from the World Wide Web: http://www.atkinsopht.com/row/rowrpage.htm (2004)
26. Baudouin, A., Hawkins, D.: A biomechanical review of factors affecting rowing performance. Br. J. Sports Med. **36**(6), 396–402 (2002)
27. Caplan, N., Gardner, T.N.: Optimization of oar blade design for improved performance in rowing. J. Sports Sci. **25**(13), 1471–1478 (2007)
28. Doyle, M.M., Lyttle, A., Elliott, B.: Comparison of force-related performance indicators between heavyweight and lightweight rowers. Sports Biomech. **9**(3), 178–192 (2010)
29. Sliasas, A., Tullis, S.: The dynamic flow behaviour of an oar blade in motion using a hydrodynamics-based shell-velocity-coupled model of a rowing stroke. Proc. Inst. Mech. Eng. Part P: J. Sports Eng. Technol. **224**(1), 9–24 (2010)
30. Coppel, A., et al.: Numerical modelling of the flow around rowing oar blades (P71). In: The Engineering of Sport 7, pp. 353–361. Springer (2008)
31. Sliasas, A., Tullis, S.: Numerical modelling of rowing blade hydrodynamics. Sports Eng. **12**(1), 31–40 (2009)
32. Kleshnev, V.: Segments contribution to the stroke length and rowing power. Higher percentage of trunk powerb help to achieve higher boat speed. Rowing Biomech. Newslett. **2**(2) (2002)
33. Coppel, A., et al.: Simulating the fluid dynamic behaviour of oar blades in competition rowing. Proc. Inst. Mech. Eng. Part P: J. Sports Eng. Technol. **224**(1), 25–35 (2010)
34. Leroyer, A., et al.: Experimental and numerical investigations of the flow around an oar blade. J. Mar. Sci. Technol. **13**(1), 1–15 (2008)
35. Caplan, N., Coppel, A., Gardner, T.: A review of propulsive mechanisms in rowing. Proc. Inst. Mech. Eng. Part P: J. Sports Eng. Technol. **224**(1), 1–8 (2010)
36. Kim, D., Gharib, M.: Characteristics of vortex formation and thrust performance in drag-based paddling propulsion. J. Exp. Biol. **214**(13), 2283–2291 (2011)
37. Lok, J., De Rond, M.: On the plasticity of institutions: containing and restoring practice breakdowns at the Cambridge University Boat Club. Acad. Manag. J.Manag. J. **56**(1), 185–207 (2013)

Chapter 4
Computational Fluid Dynamic of Rowing for Dynamic Blade Motion

4.1 Numerical Method of Blade Hydrodynamic Force

A numerical method is an approach that applies an accurate numerical approximation to solve the problem of mathematical analysis. For example, the Navier-Stoke equation is solved in fluid mechanics study using the iterative method. First, an initial guess forms successive approximations that converge with satisfactory accuracy solutions.

A fluid flow in the hydrodynamic of the oar blade is accomplished using the finite volume method. The method utilises the discretisation technique to develop an approximation of the governing equation in terms of the algebraic equation before being solved. The incompressible Navier-Stoke equation governs the equation as follows [1, 2]:

$$\frac{\partial \mathrm{v}}{\partial t} + (\mathrm{v} \cdot \nabla)\mathrm{v} = -\frac{1}{\rho}\nabla p + \nu\nabla^2\mathrm{v} + f \tag{4.1}$$

where v is the velocity of the flow, p is fluid pressure, f is an external force applied, and ν is kinematic viscosity defined as $\nu = \mu/\rho$.

Applying conservation law for an incompressible fluid, the continuity equation (Eq. 4.2) and momentum equation (Eq. 4.3) is derived as follows [3, 4]:

$$\frac{\partial \rho}{\partial t} + \nabla \cdot (\rho \mathrm{v}) = 0 \tag{4.2}$$

$$\rho\frac{\partial \mathrm{v}}{\partial t} + \rho\mathrm{v}(\mathrm{v} \cdot \nabla) = -\nabla p + \nabla \cdot \tau_{ij} + \rho g \tag{4.3}$$

where ρ is density, g is gravity acceleration and τ_{ij} is shear stress of ij component. The momentum equation can be deduced as a diffusion equation for the velocity vector, and mass conversion is used to derive the pressure.

© The Author(s), under exclusive license to Springer Nature Singapore Pte Ltd. 2025

A. A. Mohd Yusof et al., *Rowing Biomechanics and Hydrodynamics*,
Series in BioEngineering, https://doi.org/10.1007/978-981-97-7271-1_4

The solving process relies on the Reynold average Navier-Stoke model (RANS). RANS is the average time version of the Navier-Stoke equation. Reynold decomposition decomposes the time average value and fluctuating term, creating the instantaneous quality that represents rapid changing and chaotic velocity and pressure variation (Eq. 4.4) [5].

$$\mathrm{v} = \overline{\mathrm{v}} + \mathrm{v}' \tag{4.4}$$

where $\overline{\mathrm{v}}$ is average fluid velocity and v' is turbulence velocity fluctuating.

Introducing the decomposition term of two flow parameters in the incompressible Navier-Stoke equation and taking the average value of it results in the following RANS equation [6]:

$$\rho\frac{\partial \overline{\mathrm{v}_\mathrm{i}}}{\partial t} + \rho\overline{\mathrm{v}}_i\frac{\partial \overline{\mathrm{v}_j}}{\partial x_i} = -\frac{\partial \overline{p}}{\partial x_i} + \frac{\partial}{\partial x_j}\left[\mu\frac{\partial \overline{u_i}}{\partial x_j} - \rho\overline{u_i'u_j'}\right] + \rho g_j \tag{4.5}$$

where $\overline{\mathrm{v}_\mathrm{i}}$ and v_j is the mean velocity components in the xi and xj directions, $\overline{p}$ the mean pressure, $u_i'u_j'$. The term refers to the Reynold stress tensor, which describes a physical model of turbulence fluctuation. The turbulence model of k-ω is selected for the current investigation. The turbulence effect is assigned to the RANS via the eddy viscosity model equation. The model predicts the occurrence of turbulence based on the two variables: k and w. k refers to the turbulence kinetic energy while w is the specific dissipation rate.

The movement of the blade is modelled using additional reference frames. The Navier-Stoke equation for the fluid in the laboratory frame cannot be applied directly to the rotation frame of the sweep blade. Thus, these two reference frames are used. The background fluid domain is stationary, whereas the moving domain moves according to the rotational reference frame. The rotational axis is set at the oarlock, the origin (0, 0, 0). The Navier-Stoke equation is modified as the rotating reference frame equation [7, 8]:

$$\rho\frac{\partial \mathrm{v}}{\partial t} + \rho\mathrm{v}(\mathrm{v}\cdot\nabla) = -\nabla p + \rho\left[2\omega\times\mathrm{v} + \omega x(\omega\times q)\right] + \nabla\cdot\tau_{ij} + \rho g \tag{4.6}$$

where rotation rate, ω is a model using the centripetal acceleration tern and Coriolis acceleration term due to the rotation. q is the distance of the vector to the axis of blade rotation.

The Navier-Stroke equation's non-linear equation is solved implicitly using the iterative method of segregated flow solver. The momentum equation for the velocity variable (x, y, z) is first solved before moving to the following continuity equation of the pressure variable for all cells.

4.2 Computational Fluid Dynamic of the Blade

Due to the complexity and limitation of capturing the experimental data, some researchers prefer computational fluid dynamics (CFD). The method is reliable in solving issues of interest and is widely used to study fluid mechanics [9–11]. The CFD analysis was set to move linearly with a constant velocity and rotate from 0 to 180 degrees. Both quasi-static and dynamic studies of rowing propulsive provided sine and cosine graphs, respectively, for the lift and drag force [9, 12]. Coppel et al. provided a preliminary simulation around the quarter-size rectangular blade. They validated the result using Caplan et al. [13]. The computational simulation improvement was performed using a full-scale blade model to predict blade performance according to the force coefficient [14]. Besides that, Kinoshita et al. also used the CFD method to explore the development of the pressure and vortex related to the reduced frequency of the blade motion [11].

The CFD method enables quick and efficient water flow assessment around the blade. The important data that are difficult to obtain using the experimental method can be extracted easily: blade projection path, fluid flow, the force generated, water unsteadiness, and force coefficient. Blade projection path has been reported by Kleshnev using the experimental method according to the oar angle and boat velocity [15–17]. The method of path identification of CFD analysis is proven by Leroyer et al. using both data (oar and boat speed) [18]. Fluid flow is one of the important elements that can determine blade performance. Laminar and blade turbulence can be visualised, and its effect can be used to support energy conservation. Besides that, force acting can be obtained by integrating the generated pressure on the blade.

4.3 Case Study: Effect of CFD Model Simplification and Free Surface on the Blade Hydrodynamic Force in Rowing

Rowing is a propulsive sport in which the boat accelerates using the blade's hydrodynamic force. Research methodology using Computational fluid dynamic analysis (CFD) has recently attracted researchers to explore fluid flow and propulsive force generated during the drive phase. The advantage of the method is the reliability of the approach, which promises numerous benefits that could not be achieved through experimental testing. However, several proper considerations using model simplification and boundary conditions are required to obtain the best result representing the actual rowing propulsive mechanism conditions. The objective of the current study was to investigate the effect of the simplified model in CFD analysis, such as domain size, blade model, and free surface, on the hydrodynamic force. The study discovered that the domain size did not affect the blade hydrodynamic force. Meanwhile, the effect of the shaft attachment was insignificant, with a 4.67% difference, and can be neglected. A free surface is an important feature to be considered as a boundary condition to obtain realistic blade hydrodynamic force.

During the competition, the hydrodynamic force of the blade is influenced by many factors, such as the type of blade, the unsteadiness of the water, and blade motion. Several methods are used to investigate the propulsive rowing mechanism, either through the experimental or computational approach [15, 19–22]. In recent years, the computational method using computational fluid dynamics (CFD) has become necessary in sports engineering due to its reliability and low cost compared to the experimental approach. However, several basic CFD configurations must be considered to obtain accurate results and save computational time, such as domain size, blade attachment model, and free surface effect. These parameters increase the complexity of the study through the three-dimensional unsteady flow of fluid around the blade.

Previously, Leroyer et al. and Sliasas et al. investigated blade propulsion using the CFD approach [19, 23]. The blade was modelled without the shaft, and the free surface effect was considered without solid argument. Besides that, the fluid domain size is important since it may influence the blade's hydrodynamic force. Sliasas et al. used a domain size of 10 m × 10 m × 2 m, but there was no further explanation for the size selection [23]. Therefore, the present study investigated the effects of the CFD simplification of fluid domain size and shaft attachment besides free-surface boundary conditions towards blade hydrodynamic force during the drive phase. The blade was modelled based on the full-scale model of the Macon blade used by the rower.

4.4 Numerical Method

The force generated on the blade was derived based on Newton's second law in a rotating frame of incompressible flow, as shown in Eq. 4.7. The equation can be deduced into inertia force, pressure force, viscous force, Coriolis force, and external force as follows,

$$\frac{d\vec{V}}{dt} = -\frac{\nabla p}{\rho} - 2\vec{\Omega} \times \vec{V} + \vec{g} + \vec{F} \tag{4.7}$$

where $d\vec{V}/\mathrm{dt}$ is acceleration relative to the axis, $-\nabla \mathrm{p}/\rho$ is pressure gradient force, $-2\vec{\Omega} \times \vec{V}$ is Coriolis force where Ω equal to 7.929×10^{-5} rad/s, $\vec{g}$ is effective gravity equal to 9.80 $\mathrm{m/s^2}$ and $\vec{F}$ is friction.

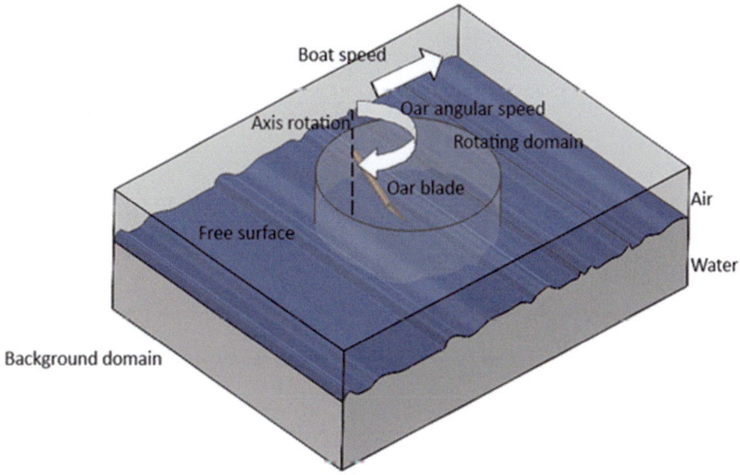

Fig. 4.1 3D schematic diagram of the CFD model

4.5 CFD for Blade Study

Computational fluid dynamic analysis was performed using Star CCM+ software (CD-adapco, version 9.06). The Reynolds Averaged Navier-Stokes (RANS) was used to predict the turbulence flow around the blade during the drive phase. The turbulent flow was calculated using the mean fluid velocity and fluctuating velocity components [24]. The analyses were executed using the control volume settings provided by the software. The scheme of the k-ω turbulence model was used for the flow to express fluid turbulence around the blade.

Three-dimensional schematic diagrams of all parameters are illustrated in Fig. 4.1. Material properties of the water (density 997.561 kg/m^3 dynamic viscosity 8.8871×10^{-4} kg/(m s)) and air (density 1.18415 kg/m^3 dynamic viscosity 1.85508×10^{-5} kg/(m s)) were applied in the simulation. However, air property was only considered when the free surface was interesting.

4.6 Boundary Condition

The blade was assigned to move in a rotational and translational motion by imitating rowing blade motion during the drive phase, as shown in Table 4.1. The axis of rotation was set at the origin. The blade was located in the rotational domain, and the communication between the rotational and background domains was done using Chimaera approaches. The flow and hydrodynamic force were obtained using the

Table 4.1 Table blade angular speed and boat translation speed were assigned to the study

Time (s)	Angular speed (rad/s)	Boat speed (m/s)
0.0	0.00000	0.000000
0.1	– 0.22639	0.002213
0.2	– 0.30294	0.010000
0.3	– 0.43947	0.020227
0.4	– 0.58489	0.099071
0.5	– 0.71171	0.210775
0.6	– 0.81033	0.346111
0.7	– 0.88322	0.496687
0.8	– 0.93912	0.654948
0.9	– 0.98736	0.814175
1.0	– 1.03201	0.968486
1.1	– 1.06611	1.112836
1.2	– 1.06597	1.243013

interpolation between the acceptor and donor mesh. The fluid domain was assumed to be a non-slip wall except for the top surface, which was assigned as an asymmetrical boundary condition. The numerical studies were solved empirically using segregated flow.

4.7 Mesh Generation in CFD

The CFD model has meshed with the Hexa element type, as shown in Fig. 4.2. Mesh was set to dense around the blade, which was 50% smaller than the mesh size for the background domain. A mesh convergent test was carried out to assess the accuracy of the CFD result and computational time. A total of 868,853 numbers of meshes were selected to apply in the following computational analysis.

Figure 4.3 shows the pressure distribution on the blade used on the blade. Macon blade was situated on an incline relative to the water-free surface. The mesh was arranged according to the fluid domain orientation. Small differences in total pressure were observed for each of these mesh numbers. These values were obtained across the dotted line on the blade between 0.0 and 0.4 m.

Generally, higher mesh numbers generated higher total pressure before they converged at 4 million mesh numbers. The percentage of error was used to obtain the appropriate number of elements by comparing the total pressure obtained with the total pressure of 4 million number elements. The mesh number of 868,853 with the lowest percentage error was selected for further analysis (refer to Table 4.2). Besides that, the pressure contour for the 868,853 number of elements was the closest pattern to the 4 million elements. The red colour pressure was consistently distributed near the blade tips compared to the other mesh elements.

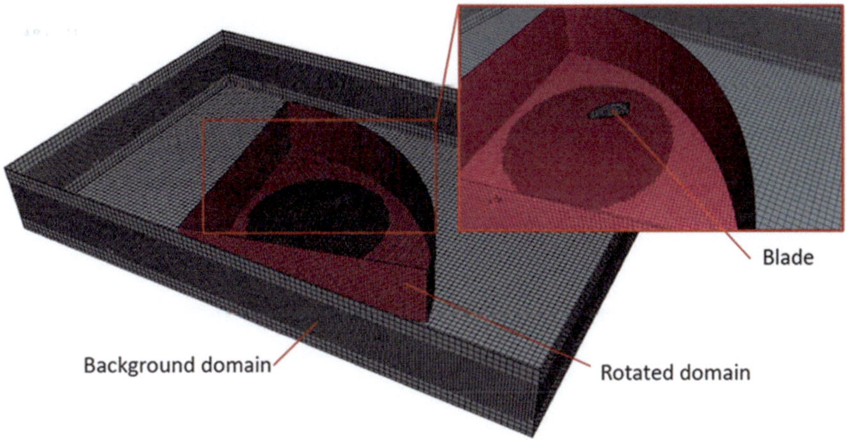

Fig. 4.2 Mesh domain used in the simulation

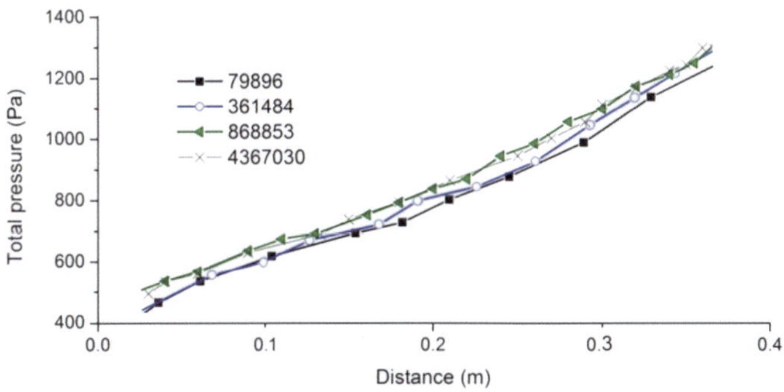

Fig. 4.3 Plotted total pressure acting on the blade at the 0.0–0.4 m distance, shown by the dotted line for the different mesh sizes

Table 4.2 Detail of convergent study

The mesh size of the rotation part (m)	The mesh size of the background part (m)	No. of mesh element	Computational time (h)	Percentage error
0.09	0.20	79,896	3.59	6.75
0.06	0.10	361,484	17.89	4.79
0.04	0.08	868,853	25.00	1.05
0.02	0.06	4,367,030	131.20	Reference

4.8 Study Parameters for CFD Simplification and Free Surface Effect

A simplified Computational Fluid Dynamics (CFD) model, used to study fluid flow related to rowing propulsion, offers several benefits, particularly useful in sports engineering and scientific applications. In this study, the accurate simplified CFD models were carried out to achieve less computational power and time than more complex and detailed simulations. This makes the study more accessible and efficient, especially when studying large-scale systems or when limited computational resources are available. The description of the CFD simplified model and boundary condition of the free surface was explained as follows.

4.8.1 Flume Size

The analysis was carried out to determine the appropriate size of the fluid domain and the effect of the wall on blade propulsion, as illustrated in Fig. 4.4a. The real size of the experimental water tank, which was 15 m × 5 m × 0.62 m, was used as a control. The blade was located near the wall of the tank to represent the blade catch condition. The length of the second model of the CFD domain was shorter than the real size, while other dimensions and blade conditions remained the same. The length of the fluid domain for the second model was 8.50 m, which was enough for the blade to rotate and translate without hitting the wall, as illustrated in Fig. 4.4b.

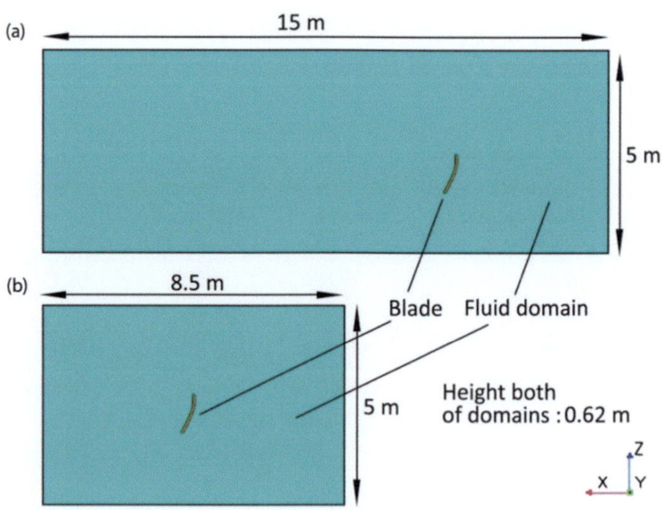

Fig. 4.4 CFD domain size **a** real size of the CFD model according to the water tank **b** the smaller size of the CFD model

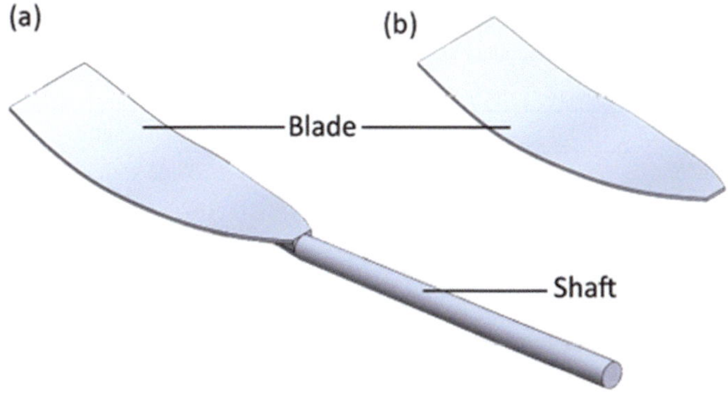

Fig. 4.5 Blade model for CFD study **a** blade model with shaft attachment. **b** Blade model without shaft attachment domain size effect

4.8.2 Shaft Attachment

The effect of the blade with and without the shaft attached was investigated to save simulation time. The blade was constructed by having the shaft attached at one end (Fig. 4.5a) and the other without the shaft, as in Fig. 4.5b.

4.8.3 Free Surface

The free surface is the boundary between water and air in which both the elements cannot pass and mix, as shown in Fig. 4.6. Free surface determines the steadiness and viscous effect of the submerged blade, affecting the magnitude of the blade's hydrodynamic force.

The study maintained the water depth for both models without a free surface and a free surface. Free surface water was introduced into the model using the Eulerian multiphase physic model according to the governing equation 4.8.

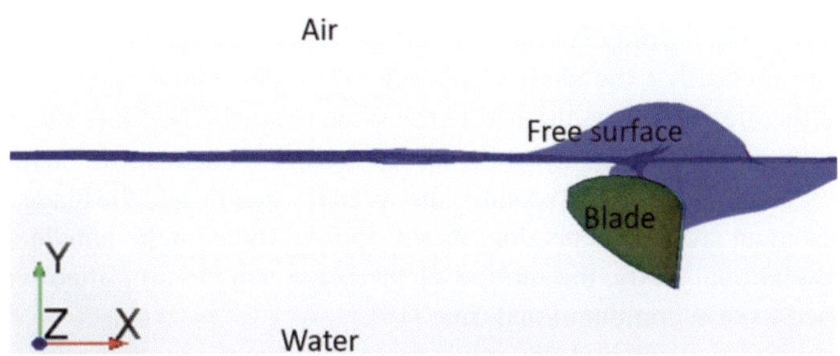

Fig. 4.6 The free surface of the water during the CFD study

Table 4.3 A detailed comparison between the actual condition of the rowing propulsive, Case 1 and simplification model and boundary conditions, Case 2

Simplified configuration	Case 1	Case 2
Domain size	15 m length	8.5 m length
Shaft attachment	Blade with shaft	Blade without shaft
Free surface	With free surface	Without free surface

$$\left[\left(v - v_{fs}\right) \cdot n\right]_{fs} = 0 \text{ or } \dot{m}_{fs} = 0 \tag{4.8}$$

In where $v - v_{fs}$ is the relative velocity between the normal component of the water velocity at the surface, where the normal velocity component of the free surface and f_s denotes the free surface. The mass flow rate, known as the kinetic condition, was simplified as zero.

The details of the studies are provided in Table 4.2. Case 1 was the actual condition of the rowing propulsion, while case 2 was the simplified model and boundary conditions (Table 4.3).

4.9 Result and Discussion of Parameters Study

The hydrodynamic force acting on the blade during the drive phase for the real and shorter fluid domains were compared, as shown in Fig. 4.7. It shows that the generated force had similar patterns and was perfectly matched. It is suggested that the domain size, especially the domain length, did not affect the blade hydrodynamic force, which could provide a similar fluid characteristic around the blade during the drive phase.

According to the Navier-Stoke equation, the hydrodynamic force generated was due to the changing momentum of the moving fluid around the blade [25]. Meanwhile, the domain size did not affect the fluid distance from the blade and can be ignored. However, since the domain size did not influence the force generated, a smaller domain with just enough space for the blade to rotate should be considered in further studies to save computational time.

Figure 4.8 shows the effect of the shaft attached to the blade on the hydrodynamic force. It was found that the shaft increased the blade hydrodynamic force up to 4.67%, and the graph was maintained in the same pattern. The shaft attached to the blade contributed little to the hydrodynamic force generated. In the actual rowing condition, the shaft is required to connect between the handle and the blade. Thus, the blade moves in an arc path depending on the applied force on the handle. However, in the computational study, the motion of the blade can be simplified without the shaft attached to save computational time [19].

The slight difference in hydrodynamic force between the models with or without a shaft attached was due to the cylindrical shape of the shaft when compared to

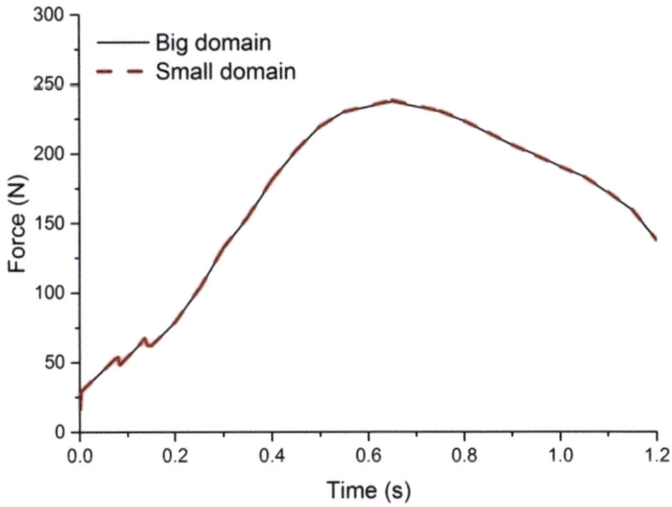

Fig. 4.7 Comparison between blade hydrodynamic force of the big and smaller CFD domain size

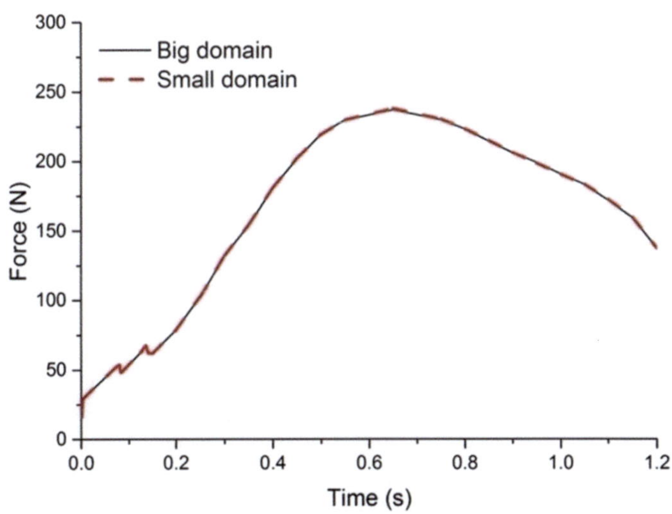

Fig. 4.8 Comparison between blade hydrodynamic force of the blade model with shaft attachment and the blade model without shaft attachment

the blade design [26]. The cylindrical shape of the shaft directed the fluid flow aerodynamically without major alterations to the fluid flow around the blade.

Figure 4.9 compares hydrodynamic force generated with two different conditions, with and without a free surface. It was found that a free surface reduced the blade hydrodynamic force by up to 10% compared to the CFD model without a free surface.

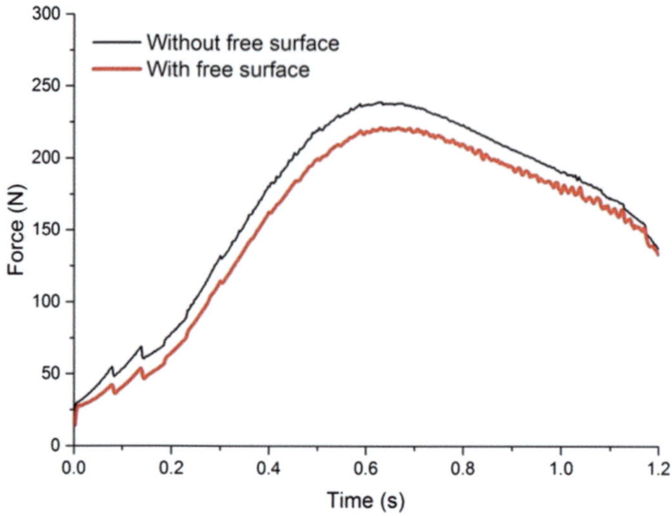

Fig. 4.9 Comparison of the blade hydrodynamic force between the CFD, which considered the free surface effect, and the CFD study, which ignored the free surface effect

The free surface is the boundary that separates two homogenous fluids: water and air. Water in the gravitational field is considered to form a free surface if unconfined from above [27]. The effect of the free surface on the row type of propulsive mechanism is significant, as Leroyer et al. reported using a simplified model and blade motion [28].

Placing the blade about 10 cm below the free surface causes higher water to become unsteady, an air pocket at the back of the blade, and waves leading to a reduction of the force generation during the drive phase [28]. Concerning this result and the results of previous studies, the free water surface should be included to obtain a realistic result representing the blade's hydrodynamics [19, 22, 23].

The performance of the rowing depends on the blade. Caplan et al. reported that the shape of the blade and its features influence the magnitude of blade hydrodynamic force. For the computational study, the computational configuration may also contribute to the force magnitude [15, 29]. Thus, it is crucial to understand the effect of the configuration of the computational study in predicting the blade hydrodynamic force. Failure to do this causes over or underprediction of the result and incorrect analysis.

The section study aimed to investigate the effects of the CFD simplified model and free surface on the hydrodynamic force of the blade. The advantages of the study are that it will help us understand the selected boundary condition and the appropriate model used. There was no significant CFD domain size and shaft attachment to the hydrodynamic force generated. Thus, domain size can be reduced, and blade attachment can be removed from the model to reduce computational costs while obtaining accurate results. However, the CFD's free surface effect must be considered to assess the blade hydrodynamics viscous effect. The study was only limited to the drive phases of the rowing stroke. The results could not be expected to remain valid outside the boundary conditions and settings used here. Thus, a more compressive rowing mechanism is required to apply to other scopes of studies.

References

1. Chacón Rebollo, T., Lewandowski, R.: Incompressible Navier–Stokes equations. In: Mathematical and Numerical Foundations of Turbulence Models and Applications, pp. 7–44. Springer, New York (2014)
2. Olshanskii, M.A., Reusken, A.: Navier-Stokes equations in rotation form: a robust multigrid solver for the velocity problem. SIAM J. Sci. Comput. **23**(5), 1683–1706 (2002)
3. Bhaskaran, R., Collins, L.: Introduction to CFD Basics. Cornell University-Sibley School of Mechanical and Aerospace Engineering (2002)
4. Anderson, J., Jr.: Governing equations of fluid dynamics. In: Computational Fluid Dynamics, pp. 15–51. Springer (2009)
5. Adrian, R., Christensen, K., Liu, Z.-C.: Analysis and interpretation of instantaneous turbulent velocity fields. Exp. Fluids **29**(3), 275–290 (2000)
6. Cowles, G., Martinelli, L.: A viscous multiblock flow solver for free-surface calculations on complex geometries. In: Proc. 22nd International Symposium on Naval Hydrodynamics, Washington, DC (1998)
7. Giga, Y., et al.: Navier-Stokes equations in a rotating frame in $\mathbb{R}^3$ with initial data nondecreasing at infinity. Hokkaido Math. J. (2), 321–364 (2006)
8. Codina, R.: Numerical solution of the incompressible Navier-Stokes equations with Coriolis forces based on the discretization of the total time derivative. J. Comput. Phys. **148**(2), 467–496 (1999)
9. Sliasas, A., Tullis, S.: Numerical modelling of rowing blade hydrodynamics. Sports Eng. **12**(1), 31–40 (2009)
10. Leroyer, A., et al.: Influence of free surface, unsteadiness and viscous effects on oar blade hydrodynamic loads. J. Sports Sci. **28**(12), 1287–1298 (2010)
11. Kinoshita, T., et al.: Rowing velocity prediction program with estimating hydrodynamic load acting on an oar blade. In: Kato, N., Kamimura, S. (eds.) Bio-mechanisms of Swimming and Flying, pp. 345–359. Springer, Japan (2008)
12. Caplan, N., Gardner, T.N.: Optimization of oar blade design for improved performance in rowing. J. Sports Sci. **25**(13), 1471–1478 (2007)
13. Coppel, A., et al.: Numerical modelling of the flow around rowing oar blades (P71). In: The Engineering of Sport 7, pp. 353–361. Springer (2008)
14. Coppel, A., et al.: Simulating the fluid dynamic behaviour of oar blades in competition rowing. Proc. Inst. Mech. Eng. Part P: J. Sports Eng. Technol. **224**(1), 25–35 (2010)
15. Caplan, N., Coppel, A., Gardner, T.: A review of propulsive mechanisms in rowing. Proc. Inst. Mech. Eng. Part P: J. Sports Eng. Technol. **224**(1), 1–8 (2010)
16. Kleshnev, V.: Propulsive efficiency of rowing. In: Proceedings of the XVII International Symposium on Biomechanics in Sports (1999)
17. Sliasas, A., Tullis, S.: The dynamic flow behaviour of an oar blade in motion using a hydrodynamics-based shell-velocity-coupled model of a rowing stroke. Proc. Inst. Mech. Eng. Part P: J. Sports Eng. Technol. **224**(1), 9–24 (2010)
18. Leroyer, A., et al.: Experimental and numerical investigations of the flow around an oar blade. J. Mar. Sci. Technol. **13**(1), 1–15 (2008)
19. Sport, R., Paintworks, D.: FISA Rule Book
20. Smith, R.: The relationship between seat movement and boat acceleration during sculling. In: ISBS-Conference Proceedings Archive (1999)

21. Barrett, R.S., Manning, J.M.: Relationships between rigging set-up, anthropometry, physical capacity, rowing kinematics and rowing performance. Sports Biomech./Int. Soc. Biomech. Sports **3**(2), 221–235 (2004)
22. Kleshnev, V.: Segments contribution to the stroke length and rower power. Higher percentage of the trunk power helps to achieve higher boat speed. Rowing Biomech. Newslett. **2**(2) (2002)
23. Steinacker, J.M., et al.: Training of rowers before world championships. Med. Sci. Sports Exerc. **30**(7), 1158–1163 (1998)
24. Baudouin, A., Hawkins, D.: A biomechanical review of factors affecting rowing performance. Br. J. Sports Med. **36**(6), 396–402 (2002)
25. Kleshnev, V.: Rowing styles. Rowing Biomech. Newslett. **6**(60) (2006)
26. Caplan, N., Gardner, T.N.: A fluid dynamic investigation of the Big Blade and Macon oar blade designs in rowing propulsion. J. Sports Sci. **25**(6), 643–650 (2007)
27. Coppel, A., Gardner, T., Caplan, N., Hargreaves, D.: Oar blade force coefficients and a mathematical model rowing. In: 27 International Conference on Biomechanics in Sports (2009)
28. Perkin, H.: Teaching the nations how to play: sport and society in the British Empire and Commonwealth. Int. J. Hist. Sport **6**(2), 145–155 (1989)
29. Greidanus, A., Delfos, R., Westerweel, J.: Rowing faster by surface treatment. Procedia Eng. **34**, 361–366 (2012)

Chapter 5
Computational Fluid Dynamic of Rowing Using Fluid-Structure Interaction

5.1 Introduction to Fluid-Structure Interaction

Each rowing athlete aims to generate higher propulsive force and move faster to the competition's finishing line. Significant improvement in winning time over the years, starting from 1912 to 2008, for the Olympic Games, is achieved with the regression curve of r = 0.62, which shows steady upgrading until the present day [1]. An improved recorded time is very interesting as it contributes to implementing the advanced training method, improving fitness levels, and using the latest rowing equipment [1–3].

Rowing sport is categorized as a repeated sport type, which repeats the several moving phases for one cycle. The phases are divided into the catch, drive, finish and recovery [1, 3, 4]. The oar is pulled during the drive to generate the propulsion force and accelerate the boat. The boat's movement combines the boat's linear motion and the angular velocity of the oar blade about the oarlock with the driving angle between 35° and 135° [4].

Water flows parallel to the blade surface at a low angle of attack (~ 35°), resulting in higher lift than drag force [5, 6]. The working mechanism of the blade can be represented as a hydrofoil, where there is a pressure difference on the front and back surface of the blade, resulting in the generation of lift and drag force. The lift force acting on the blade increases from the lowest angle to 45°, whereas the drag force decreases [5]. However, from 45° to 90°, the drag force increases and the lift force decreases. These situations then keep repeating until the finishing line.

The working mechanism of the blade can be represented in the same way as a hydrofoil. Assuming the fluid is incompressible, steady, irrational, and the body forces are conservative. Furthermore, there are sinks or sources of energy along the streamline, which is described in Bernoulli's equation [7]:

$$\frac{P_1}{\rho} + \frac{V_1^2}{2} + gz_1 = \frac{P_2}{\rho} + \frac{V_2^2}{2} + gz_2 = Constant \tag{5.1}$$

© The Author(s), under exclusive license to Springer Nature Singapore Pte Ltd. 2025

A. A. Mohd Yusof et al., *Rowing Biomechanics and Hydrodynamics*,
Series in BioEngineering, https://doi.org/10.1007/978-981-97-7271-1_5

where P, V, ρ, g and z are pressure, fluid velocity, gravity force and height of the point. The pressure and height are ignored since their effects are minor compared to those due to kinematics and pressure. Considering the front and back surfaces of the blade, the final equation suits Eq. 5.2. Hence, as the velocity of the back surface is slower than the velocity of the front surface, to satisfy the equation, the back pressure must be higher than the front pressure.

$$\frac{P_{back}}{\rho} + \frac{V_{back}^2}{2} = \frac{P_{front}}{\rho} + \frac{V_{front}^2}{2} = Constant \tag{5.2}$$

The fluid dynamics of rowing have been the subject of extensive research, aiming to improve performance and understand the sport's physics. One approach to studying rowing is through the use of fluid-structure interaction (FSI) methods, which combine computational fluid dynamics (CFD) with finite element analysis (FEA) to simulate the interaction between the fluid and the structure of the boat and oars.

FSI simulations can provide insight into the complex flow patterns and forces that occur during rowing, as well as the response of the boat and oars to these forces. By studying the FSI of rowing, researchers can identify areas where performance can be improved, such as reducing drag or optimizing the motion of the oars. This can help athletes and coaches make informed decisions about training and equipment, leading to better performance on the water. By simulating the interaction between the boat, oars, and water, we will better understand rowing mechanics and identify areas where improvements can be made.

In this study, an analytical processing method of calculating fluid-structure interaction (FSI) was developed to investigate the parameter of blade design while considering the effect of the net pressure generated, which is related to the performance of rowing sports. Studying the interaction between fluid flow and solid structure mechanics is an important step toward understanding the relationship between the athlete's ability, structure and fluid for improving hydrodynamic propulsion, especially in rowing sport.

5.2 Computational Simulation of Quasi-static Investigation

In this study, Solid and fluid domains were modelled according to CFD and FEA settings. The study included the solid domain to represent blades submerged in the fluid domain, as in Fig. 5.1. The first two blade models were Macon and Cleaver. The other two models were the simplified model of the Macon and Cleaver, which maintained the shaft attachment position and projected the area based on the original model. Table 5.1 shows the blade's dimension and projection area, whereas Table 5.2 shows the important data on the water flume used in the investigation.

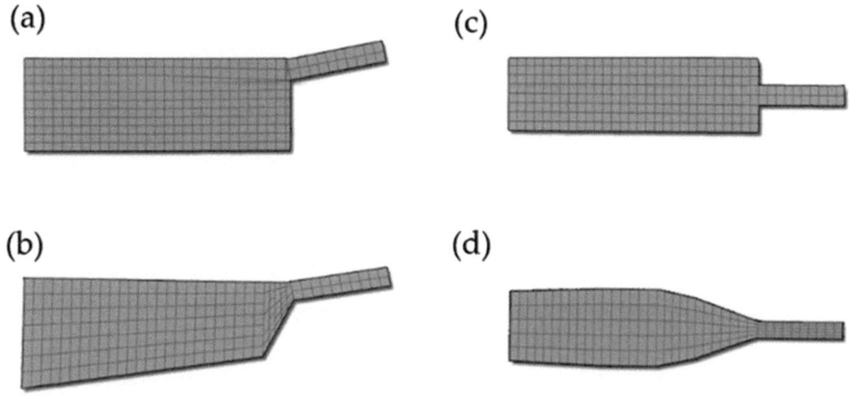

Fig. 5.1 Blade model for the simulation after meshing **a** simplified cleaver model **b** cleaver model **c** simplified Macon **d** Macon model

Table 5.1 The dimension and projection area of the blade

Blade model	Detail		
	Projection area (mm^2)	Length (mm)	Width (mm)
Cleaver	119,041	550	270
Simplified cleaver	111,769	550	200
Macon	97,234	579	200
Simplified Macon	97,440	580	168

Table 5.2 The important data of the water flume

Flume	Detail
Width (mm)	900
Length (mm)	1500
Depth (mm)	500
Fluid velocity (m/s)	2.5

Blade and flume models were constructed using Solidwork software and imported into ABAQUS for positioning and performing simulation. The Fluid-structure interaction (FSI) analysis combines CFD and FEA simultaneously while looking for results. In general, FSI is iteratively implicit, where the load development is repeated by transferring the value at the interface of the domains. However, since this problem involves large deformation due to the impact of the fluid, a dynamic explicit step was used instead of an implicit one.

Both domains were meshed with different mesh sizes and element types and simulated under different conditions and algorithms. The size of the blades was in the range of 168–579 mm, and each small deformation had to be caught. Hence, the

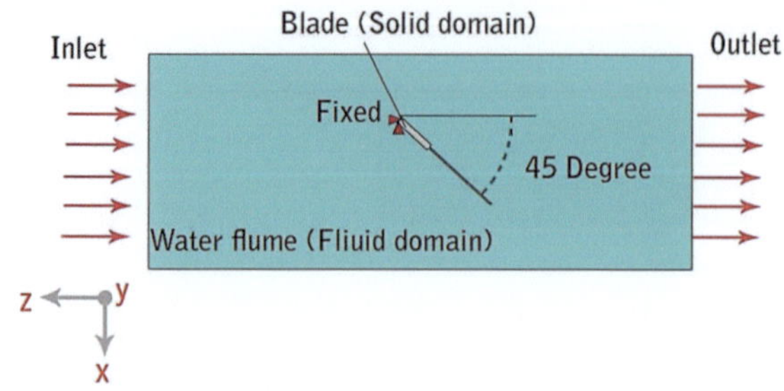

Fig. 5.2 Boundary condition for the FSI

best size of the element was 20 mm, whereas the flume size was considered big, and the interface between the blade and flume was really important to ensure detailed information was correctly transferred. The appropriate sizes were 50 mm, and the interface's meshing size was identical to what was assigned to the blade.

Solid and fluid domains were assumed to have homogeneous properties since the study looked for the effect of the blade design. The density of 1E−009 and dynamic viscosity of 7.89E−007 were used for water. For the blade, the density of fiberglass material with the value of 2E−009 was assigned to the structure. In the mechanical aspect, two important mechanical properties of fibreglass, Young's modulus and Poisson's ratio, were selected at 10,000 MPa and 0.3, respectively.

Referring to Fig. 5.2, water velocity was set to 2.5 m/s at the inlet. At the outlet opening, static zero pressure was preferred. Hence, the final result is the generated pressure, which is generally due to water velocity when it passes through the static blade structure. Blade structures were held by fitting the stick at 45° relative to the flume wall. In this position, the lift force generated was at its maximum, and the drag force was minimized. The side wall was modelled as a non-slip wall; therefore, the blade was assumed to be fully submerged, and the fluid-structure interaction happened at the whole blade surface. The simulation was carried out for four designs, as shown in Fig. 5.2, to investigate the parameters' effect on fluid-structure interaction.

5.3 Results and Discussion

A preliminary study using simulation software from the previous simulation work has been launched to generate the result according to FSI. The analysis was simplified as quasi-static, where the blade was held at 45°. It looks for the deformation of the blade structure and the net pressure the moving water produces to evaluate the model performance as the hydrofoil.

Figure 5.3a illustrates the arrows representing the vector direction of the moving water from the inlet to the outlet. In the early stage, the water was parallel to the wall of the flume and moved at 2.5 m/s. However, as the moving water close to the blade, the velocity was reduced slightly, averaging 2.2 m/s. The flow was then separated into four directions: top, bottom, left and right. Once water hits the blade, the passed velocity gets even lower, with an average of 1.9 m/s and embraces a more extensive blade area. Depending on the design of the blade, the separated water was combined and flowed in the same direction as the initial state toward the outlet.

Referring to the Bernoulli equation, pressure is inversely proportional to the fluid velocity. Therefore, the front surface of the generated pressure was low as the speed was high. However, the pressure at the back surface increased with low water velocity. Different blade designs contribute to the velocity profiles and create different net pressures, as shown in Fig. 5.3b.

To investigate the function of each parameter related to other blade design, the comparison was done by dividing the blade into two categories. The first group was the Cleaver design, as it varies in the method of blade-shaft connection, as well as in the asymmetrical and projection areas. In contrast, the second group was classified as the Macon design. The main characteristic of the Macon blade was that it was

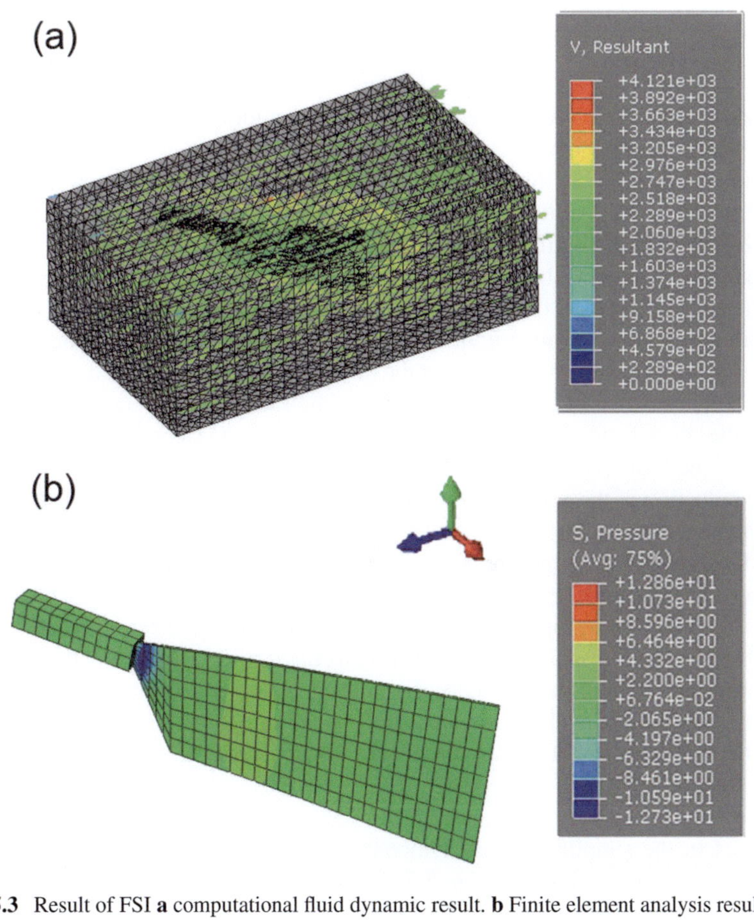

Fig. 5.3 Result of FSI **a** computational fluid dynamic result. **b** Finite element analysis result

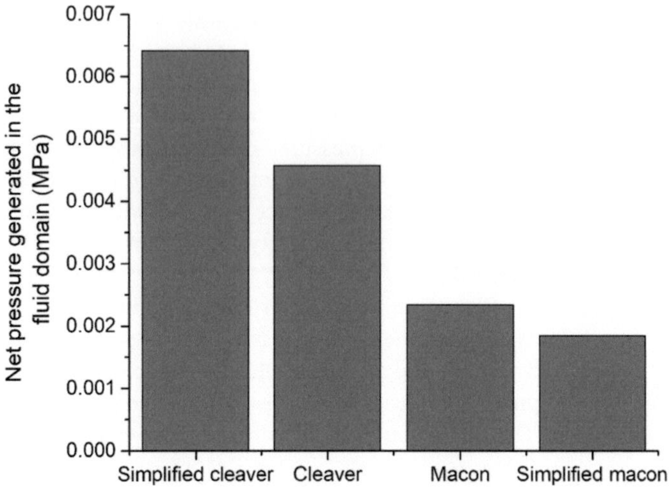

Fig. 5.4 Blade design versus net pressure generated for the fluid

symmetrical, and the shaft was connected to the middle edge of the blade surface and lower projection area.

Referring to Fig. 5.4, the highest net pressure generated when the water moved from the inlet to the outlet happened to the simplified Cleaver, followed by the Cleaver, the Macon, and the simplified Macon. The main reason that contributed to the higher net force was that the projection area of the Cleaver was 22% bigger than the Macon blade. Hence, the blade can hold more water to interact and act as a solid anchor to push the boat forward. In the design aspect between the simplified Cleaver and the Cleaver, the result showed that the design adjustment on the side of the blade reduced the net force by 29% lower than the simplified model.

On the other hand, in the case of the second group (Macon blade design), the design of the Macon blade performed better than the simplified Macon blade, with a 26% reduction of the generated pressure. This can be explained by using the blade design. Since it is designed symmetrical, the features contribute to the higher aspect ratio, leading to lower pressure generation. The aspect ratio in the study was used to describe the blade length ratio to average width. Unlike the wing aspect ratio, the blade works as the drag-induced, where the wider the blade, the deeper it submerges.

When the water hits the blade structure, deformation occurs, bending the structure backwards. As shown in Fig. 5.3b, there are negative and positive values of the pressure experienced by the blade, which means the structure is bent at two locations, near the middle of the blade and the connection point of the shaft and blade. The same situation applies to all tested blade models.

Similar to the water's net-generated pressure, the structure's highest pressure happened in the simplified Cleaver model, followed by the Cleaver, Macon, and simplified Macon, as shown in Fig. 5.5. The difference in pressure experienced on the blade between the simplified Cleaver and the Cleaver was 42 and 23% between

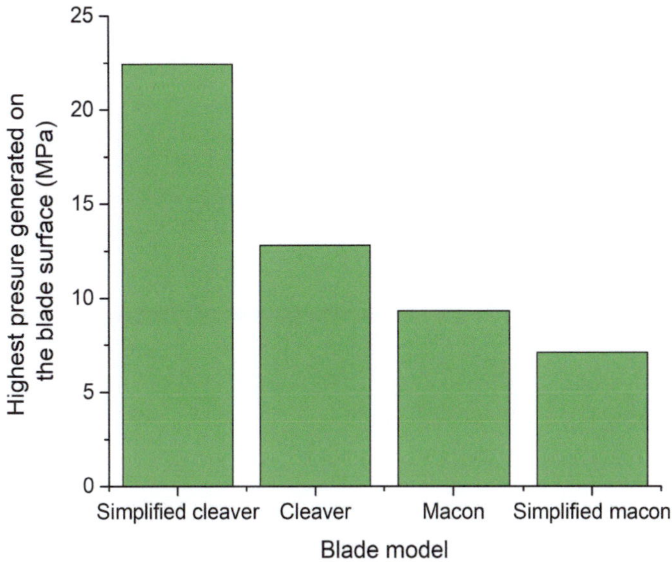

Fig. 5.5 Blade design versus highest pressure generated for the structure

Macon and the simplified Macon. This happened as the generated pressure from the water was transferred to the blade structure.

References

1. Caplan, N., Coppel, A., Gardner, T.: A review of propulsive mechanisms in rowing. Proc. Inst. Mech. Eng. Part P: J. Sports Eng. Technol. **224**(1), 1–8 (2010)
2. Greidanus, A., Delfos, R., Westerweel, J.: Rowing faster by surface treatment. Procedia Eng. **34**, 361–366 (2012)
3. Caldwell, J.S., McNair, P.J., Williams, M.: The effects of repetitive motion on lumbar flexion and erector spinae muscle activity in rowers. Clin. Biomech. **18**(8), 704–711 (2003)
4. Baudouin, A., Hawkins, D.: A biomechanical review of factors affecting rowing performance. Br. J. Sports Med. **36**(6), 396–402 (2002)
5. Sliasas, A., Tullis, S.: Numerical modelling of rowing blade hydrodynamics. Sports Eng. **12**(1), 31–40 (2009)
6. Coppel, A., et al.: Numerical modelling of the flow around rowing oar blades (P71). In: The Engineering of Sport 7, pp. 353–361. Springer (2008)
7. Cengel, Y., Cimbala, J.: Ebook: Fluid Mechanics Fundamentals and Applications (SI Units). McGraw Hill (2013)

Chapter 6
Computational Fluid Dynamic of Rowing for Quasi-static and Dynamic Due to Rower Force Profile

6.1 Introduction to the Quasi-static and Dynamic Rotation of the Blade

Stroke, in the scope of this study, refers to one of the rowing techniques in which a rower applies musculoskeletal force to the oar's handle. As a result, the oar rotated at a certain angular velocity and accelerated the boat. During the stroke, the oar blade acts as the intermediate component to deliver rower load from handle to blade and generate the corresponding hydrodynamic force.

The musculoskeletal force applied on the oar's handle during the stroke can be visualized as the force profile. To complete one force profile, the leg generally contributes 46.4% of the power, 30.9% from the hand, and 22.7% from the trunk, which can be classified as rowing biomechanics. Baudouin mentioned in his paper that biomechanics affect rowing performance. From the report, he suggests that subtle biomechanics may contribute to the main factor of performance. At the same time, Smith interprets the biomechanics of the rower as a force angle graph, which can be used to evaluate biomechanics that relate to the rower's body segment. The force profile can be grouped into three main groups based on the graph pattern: (1) jump the catch, (2) Over-emphasis finish, and (3) balanced force.

Several methods have been described to improve rowing performance in rower biomechanics. The first aspect is anthropometry, especially in the height of the rower [1]. In agreement with the previous studies and as has been reported in some of the journals, the height factor significantly affects performance, which is one of the roles of extending stroke time and producing more force [2–4]. On the same topic of anthropometry, the rower's higher mass can also increase biomechanical force. However, at the same time, the additional mass affects the boat by increasing the drag force. Besides that, stroke rate is an important aspect that was considered by all of the selected papers. Increasing the stroke rate directly leads to increased biomechanical force, which speeds up the boat's velocity [3, 5–8].

© The Author(s), under exclusive license to Springer Nature Singapore Pte Ltd. 2025
A. A. Mohd Yusof et al., *Rowing Biomechanics and Hydrodynamics*,
Series in BioEngineering, https://doi.org/10.1007/978-981-97-7271-1_6

The oar blade projection during stroke is shown in Fig. 6.1. The oar was reported as a lever that can be used to magnify the force using inboard to outboard shaft length. There are three main elements related to the lever: fulcrum, load and effort. However, two lever classes are involved to ensure the oar works correctly. The first is a first-class lever where the fulcrum, FO, acts in the middle, effort acts in the handle, FH and load, and FP acts in the blade. The working principle of this lever causes the oar to rotate freely by pivoting at the oarlock. The second one is the second class lever, where the fulcrum acts at the blade, FP as the imaginary fulcrum, and effort, FH still act as the handle and load, and FO at the middle of the oar to accelerate the rowing boat and rower. The working principle of this lever causes the whole rowing system to move forward. Due to the rotation and translation motion acting on one of the oars, the blade then moves in a complicated path, as in Fig. 6.1.

As the oar is rowed and accelerates the boat, the rowing system has three forces: propulsive force acting on the blade, drag force from water and air acting as resistance to the movement, and the net force accelerating the boat. Only propulsive force acting on the oar contributes to boat acceleration. Thus, it is important to understand the fluid mechanics of the blade as it could provide a deeper insight into the improvement. Specifically, the stroke for the drive phase can be separated into six sections. The first two sections are the most important parts attributed to a higher efficiency section, generating a more significant propulsive force. In addition, due to this fact, most rowers tend to increase force rapidly and then achieve the peak force in the early stage of the drive. Once the blades move from catch to an angle lower than 90°, the pressure difference increases rapidly, leading to propulsive force and thrusting the human-oar-boat system [9].

In particular, the blade's movement can be separated into two components: rotational movement as the blade travels outward away from the boat body at the early stroke. After passing 90° of the stroke angle, the blade moves inward toward the boat body to complete the stroke of a circular pathway. Due to the movement of the blade

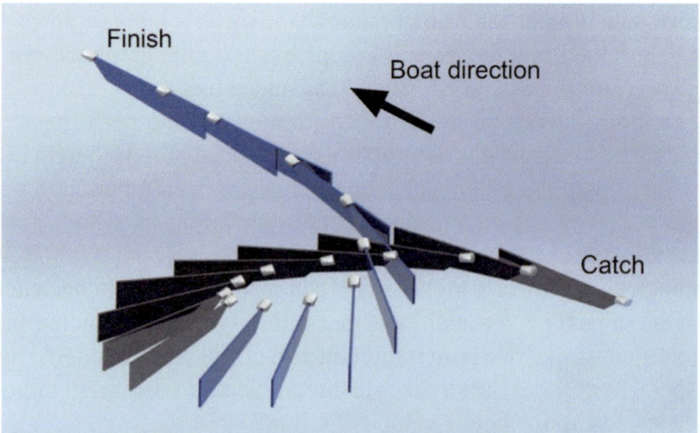

Fig. 6.1 Blade path due to rotational and translation motion adapted from [10]

in the water, hydrodynamic force, which is the combination of the lift and drag force, is generated. The lift and drag equation shows that the blade's velocity contributes to hydrodynamic propulsive force. To row faster, increasing the propulsive force and reducing the drag force are important to consider. This complex movement of the blade and its interaction with water is an interesting topic to study as it combines the hydrodynamic blade and the biomechanics of the rower.

$$F_l = \frac{C_l \rho V^2 A_o}{2} \quad \text{and} \quad F_d = \frac{C_d \rho V^2 A_o}{2} \tag{6.1}$$

Currently, the rowing problem can be studied using computational study. The method's reliability in solving the interesting issue is undoubted and widely used to study fluid mechanics. Fluids around the blade have been successfully modelled and validated by comparing them with experimental data by Leroyer. The blade was moved with linear motion constant velocity and rotated from 0° to 180°. It was found that using both qua-static and dynamic studies of rowing propulsive provided sin and cosine pattern graphs, respectively, for lift and drag force. Despite providing deeper insight regarding the propulsive mechanism of the rowing, unfortunately, most of the reported simulation works did not include the effect of rower biomechanics on blade movement.

This study presented the computational simulation of the sweep blade by considering the biomechanics stroke of the rower, which allows for the simulation of the rowing propulsive mechanic based on the rower stroke. Through the investigation, the effectiveness of each type of rower force profile can be evaluated, and the most efficient force can be suggested according to the highest average.

6.2 Computational Simulation Method for Hydrodynamic of the Blade

The study was done using the computational simulation method. Star Ccm software was used to access the hydrodynamic force of the rowing propulsive mechanics under unsteady conditions of incompressible turbulence flow. In the study, a numerical simulation of the flow around the blade was carried out at the drive phase, and the corresponding hydrodynamic force was recorded. The details related to the numerical method of the current study are explained in Sect. 4.1. At the end of the study, three rower force profiles with specific features were selected and accessed to evaluate the corresponding hydrodynamic toward performance.

6.3 Hydrodynamic of the Blade Under Quasi-static Investigation

The study was done using the computational simulation method. STAR-CCM+ software was used to access the hydrodynamic force of the rowing propulsive mechanics under unsteady conditions of incompressible turbulence flow. In the study, a numerical simulation of the flow around the blade was carried out at the drive phase, and the corresponding hydrodynamic force was recorded. At the end of the study, three rower force profiles with specific features were selected and accessed to evaluate the corresponding hydrodynamic toward performance.

The quasi-static in this scope of the study refers to an approximation or simplification of the fluid flow analysis by assuming that the motion of the blades was too slow and did not significantly affect the fluid flow patterns during a small time step. The blade was positioned in a series of angles between 0° and 180° to achieve the condition. The fluid flow around the blade was solved using the blades' current position, making the simulation more computationally tractable.

6.4 Computational Setup of Quasi-static Investigation for Rectangular Blade

This section was carried out for validation purposes based on the previous experiment. In the investigation done by Coppel, a quarter-size blade model was used. The blade was held close to the water flume's surface. With a speed of 0.75 m/s, water hits the blade surface and generates the propulsive force. Using the strain gauge at a particular configuration, which the strain gauge placed on the blade's shaft, the resulting bending strain was captured and recorded.

The blade was modelled according to experiment data, and a rectangular blade shape with a projection area of 0.00773 m^2 was used. The size was the quarter scale of the commercialized blade type used in competition. The model was reconstructed using Solidworks design software before being transferred into STAR-CCM+ software for the simulation steps.

Regarding the boundary condition, as in Fig. 6.2, one side of the flume was assigned as the input with 0.75 m/s water flow, whereas the other was assigned with zero pressure output. At the top of the flume, a symmetrical boundary condition was set to replicate open channel properties. At the same time, walls with non-slip properties were assigned to the rest of the flume side. The initial kinetic parameters corresponding to the experimental data were applied to the fluid domain, which was the same as the input flow. In this case, water was assumed to flow in moving boat condition before the blade was positioned in the hitting position. Seven different angles were chosen for investigation, where the value was increased from 0° to 180°.

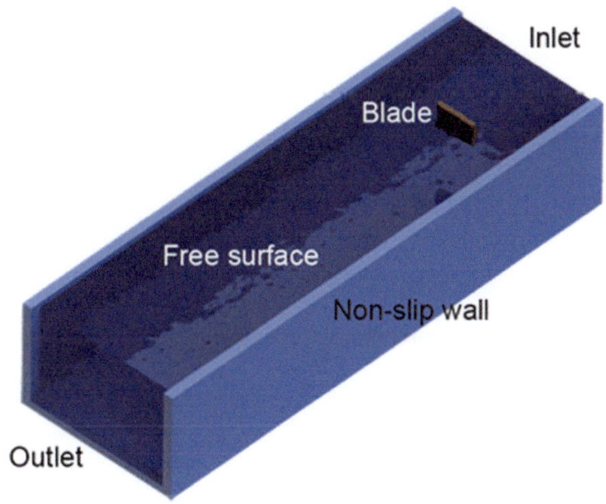

Fig. 6.2 Schematic diagram of propulsive mechanics study of rectangular blade

Normal and shear forces were obtained by integrating the normal and shear stress over the blade area. Then, force at a specific blade position was subtitled into the equation below to resolve as lift and drag force where Fn and Ft were normal force and shear force, respectively, and θ was the angle between blade and boat acceleration direction.

$$F_d = F_n \sin\theta - F_t \cos\theta \quad and \quad F_l = F_t \sin\theta + F_n \cos\theta \tag{6.2}$$

The fluid model was meshed using hexa elements or trimmed cell meshes provided by the software. The total number of elements and mesh arrangement depends on the default size assigned to the meshing element. To get the most reliable result of this investigation, as the final result depends most on the assumption that simplified the solution, it is important to confirm that the result is independent of the mesh density. By doing this, physical parameters and numerical models will change based on the manipulation data. The way to confirm the convergence of the obtained data was by looking at the value of the total normal force acting on the blade surface. This parameter was selected as it is the main contribution to the lift and drag force besides the shear force.

Figure 6.3 shows the normal force value that acted on the blade surface versus the number of elements for the fluid domain. The value of lift that acted on the blade surface decreased exponentially as the number of elements increased until 2,170,000 elements for 0.0044 m element size before converging. The obtained data was used as the baseline case for the investigation to save the simulation time at the most suitable element number.

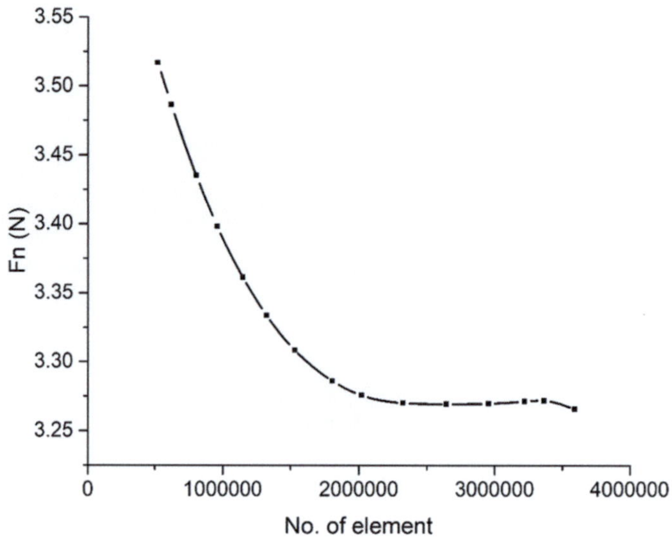

Fig. 6.3 Convergence study of the fluid domain for CFD

6.5 Validation of the Computational Result

To ensure the simulation gives a highly reliable result as the experimental, the first step is to validate the computational study via comparison with the experimental data from Caplan and Grander's study, as in Figs. 6.4 and 6.5. Better simulation results have a small percentage of error and the same pattern as the referring data. Besides that, the effectiveness of the current CFD was also important to consider: the previous CFD result of a similar study was compared for improvement purposes.

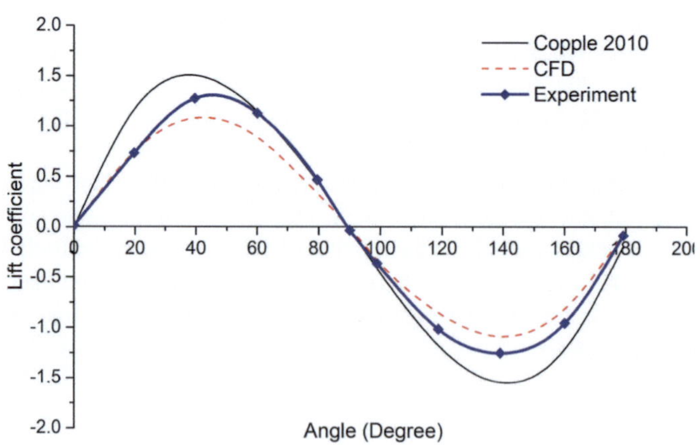

Fig. 6.4 Comparison of the drag coefficient

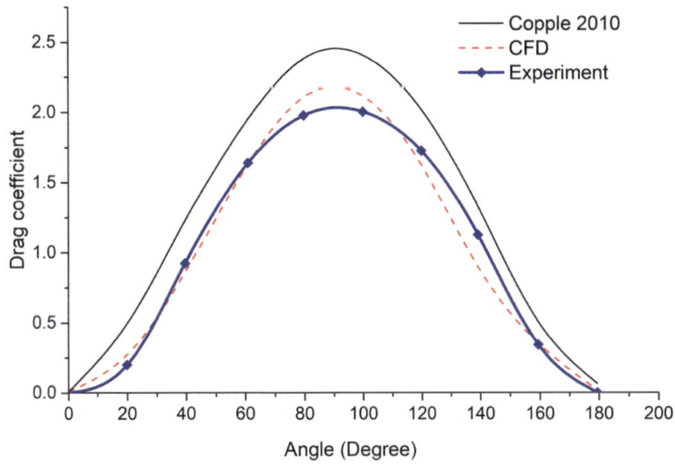

Fig. 6.5 Comparison of the lift coefficient

The comparison between the experimental study and the CFD simulation by Coppel and the current CFD simulation is shown in Fig. 6.5. The computed result of the lift and drag coefficient is compared with the experimental study and simulation of the flume test. More importantly, this result enhances the previous CFD by achieving a lower percentage error. The percentages of errors were obtained based on the area under the graph where the experimental study becomes the standard reference. The current research enhances the simulation result with 6.58 and 0.69% errors for lift and drag compared to 33.65 and 18.75% errors obtained by Coppel. It means that the computational study in this research reproduces almost the same circumstances as actual blade fluid mechanics and realistically estimates the force acting on the blade's surface.

The improvement in CFD results is influenced by the introduction of implicit unsteady setting into the fluid flow computation. On the other hand, deviation of the simulation result from the experiment happened at the peak result either for drag or lift coefficient. The drag coefficient had positive and negative peaks, and the simulation result was lower. Meanwhile, there was only one peak for the drag coefficient, and the simulation result was higher.

6.6 Computational Setup of Quasi-static Investigation for Macon Blade

The CFD was validated with a better agreement, so the study used the Macon model. The study used two sizes of Macon, as in Table 6.1. The first quarter scale model and the next was the full-scale model. The quarter scale was selected to compare with the previous rectangular blade model, while the full scale was the rower's real size. The blade was positioned at angle α, 12°, referring to the water surface as in Fig. 6.6. The blade was placed near the water surface and replicated in real condition by setting the oar in this position. In this study, the shaft was considered attached to the blade. The total length of the shaft was 2.11 m. However, only some parts submerged in the water were modelled.

Following the same physic setting and boundary condition of the flat rectangular blade, the model was replaced with a quarter of the size of a Macon. In this simulation, the oar shaft was included in the investigation to represent the propulsive condition of the actual blade size, as in Fig. 6.7. Based on the preliminary study of the shaft effect toward propulsive force, it was found that the shaft generated 4.67%, which was insignificant besides the mentioned result of blade investigation in Chap. 4. Due to the result, it was concluded that the shaft can be ignored as the blade made a significant contribution to the propulsion.

For the study, water was set to flow at 0.75 m/s and hit the blade at the angle from 0° to 180° before changing to 5 m/s for the full-size model, as in Fig. 6.6. Flow separation was expected to happen, which caused the highest pressure distribution on the front surface of the blade. The pressure was then distributed on the blade, especially for the 45-degree position. This position causes the blade's tip to be positioned onward, and the shaft causes the highest pressure region at the tip. Water flows from the front to the back surface with a uniform streamline without flow separation.

Table 6.1 Dimension of the blade used in the CFD study

Macon blade size	Length (m)	Width (m)	Projection area (m^2)
Quarter size	0.154	0.05	0.007
Full size	0.600	0.196	0.101

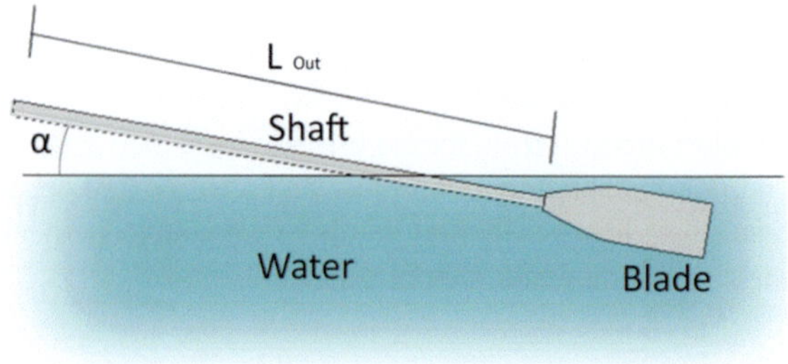

Fig. 6.6 Oar positioning for the CFD study

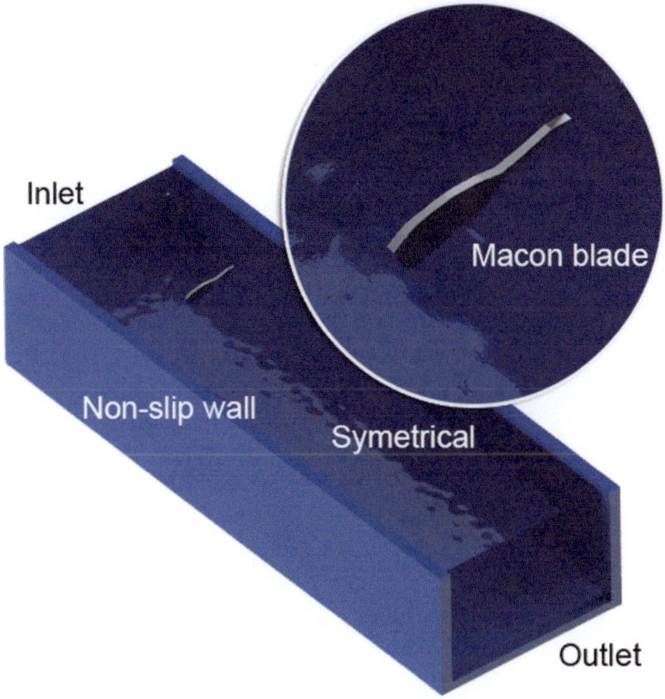

Fig. 6.7 Boundary condition of the CFD model used in the simulation

6.7 Computational Study of Dynamic Blade Rotational Investigation Based on the Rower Force Profile

Dynamic or unsteady behaviours in computational fluid dynamics (CFD) mean analysing fluid flow phenomena that change over time during the stroke. The setting considers a rower stroke that rows the blade to generate hydrodynamic force. Studying rowing stroke involves analysing the fluid flow phenomena that occur as a rower's oar blade moves through the water during a rowing stroke. This type of analysis is crucial for understanding how the rowing blade interacts with the surrounding fluid and how it generates forces and thrust during the stroke.

6.7.1 Blade Angular Angle According to Rower Force Profile

Force profile data to represent rower's biomechanics was obtained from Pilgeram's study. In the study, four strain gauges were attached to the inboard of the oar to measure the applied force and the potentiometer to capture the rotation angle. The profile graph was then plotted as in Fig. 6.8. An interesting aspect regarding this rower force profile which is that they can be divided into three groups: (1) jumping the catch, (2) over-emphasized finish, and (3) balance force [4]. Each profile has its own blade's kinematic, which reflects how the blade is swept.

Work was done once force was applied to rotate the oar at certain angles. The amount of work can be measured by the total up area under the graph. The larger and longer forces applied to rotate the oar, the more work was done. Work was then converted into kinetic energy to extract angular velocity. The kinetic energy calculation used a second-moment inertia of 6.6 kg m^2 for a sweep oar [11].

The equation can be described as follows (Table 6.2):

$$Work = \tau\theta = \frac{1}{2}I\omega^2 \tag{6.3}$$

Figure 6.9 shows the angular velocity graph of three force profiles extracted. Each graph contains the information that represents the oar kinematic. The angular velocity patterns of each graph differ from each other and depend on the force profile.

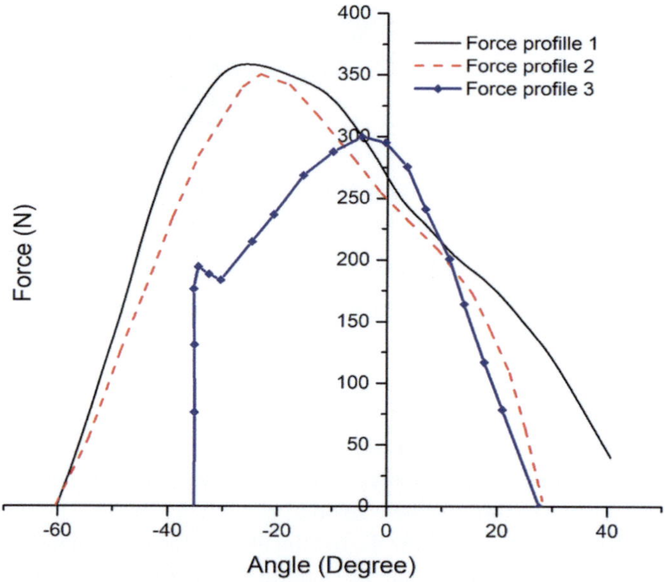

Fig. 6.8 Rower's force profile that applied on the oar handle

Table 6.2 Angular rotation detail information

Velocity profile	Total stroke time (s)	Total stroke angle (°)	Max angular velocity (rad/s)
Profile 1	0.44	62	3.71
Profile 2	0.62	88	3.99
Profile 3	0.67	100	4.07

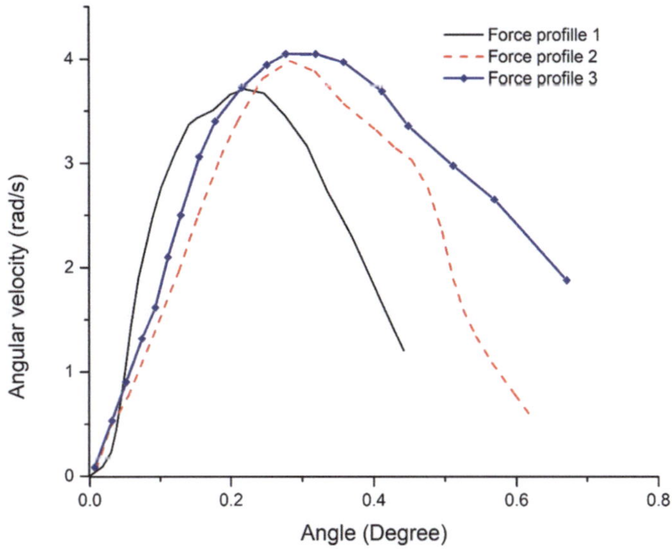

Fig. 6.9 Blade angular rotation due to rower force at the oar's handle

First profile 1 shows the angular velocity increase slowly at the first 0.02 s. However, the magnitude of the force increases rapidly, which can be interpreted as applying excessive power suddenly with a smaller oar rotation. After reaching a certain period, the boat starts to move, and the angular velocity of the oar then increases gradually before it reduces as the stroke reaches the maximum limit. It is important to note the balance between the increasing and decreasing pattern of the force profile and angular velocity of Force Profile 2, where the profile increased to the peak at the middle of the drive before reducing toward the finish. The profile growth can be elucidated as the force applied by the blade automatically moves at the corresponding angular velocity until the maximum speed. When it reaches this level, the force decreases until the finishing phase. Besides that, Force Profile 3 shows that the maximum force was maintained at a certain period to extend the supply of power to drive the boat. This angular velocity graph modified the previous pattern by keeping the peak at a specific value before gradually decreasing.

6.8 Computational Setup of Dynamic Blade Rotational Condition for Macon Blade

The blade was rotated with specific angular velocity to reflect the drive phase of the rowing stroke. While the blade was rotated simultaneously, water flowed from the inlet to the outlet at 5 m/s flow velocity. Use the same physic setting as the quasi-static condition as it was considered an unsteady flow condition with the K-Omega

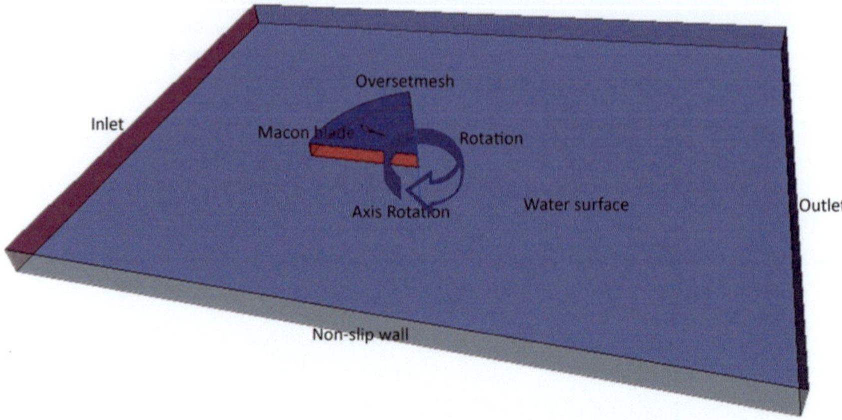

Fig. 6.10 Boundary condition for the dynamic rotation study due to rower force profile

turbulence model. At first, the flat Macon is rotated using an angular velocity profile and then replaced with the curved Macon blade. The investigation on the curved Macon is important to replicate hydrodynamic force on the actual blade model.

The overset mesh uses overlapping grids to model the blade's motion. The configuration comprises the overset grid and the background mesh, as in Fig. 6.10. The outer surfaces of the subdomain are set as overset mesh. The background then surrounds the overset mesh and couples at the overlapping interfaces. The cells at the overlapping interface between the overset and background are interpolated using a linear option to obtain the fluid variables.

The blade curvature effect is an important design element that must be included in the study. The design element cannot simply be ignored and needs to be included in this simulation study. Compared to a flat Macon blade, the curved design was introduced to the middle of the blade to make it bent, as in Fig. 6.11. The blade curve is expected to grab more fluid and alter the flow to generate higher hydrodynamic force. In this section, the mentioned force profiles (Fig. 6.7) matched the curved blade to access generated hydrodynamic forces for better comparison.

6.9 Result of the Computational Study

The computational fluid dynamic simulation was carried out in four situations that differ in the size of the blade (quarter and full scale), hydrodynamic condition (quasi-static and dynamic), and blade parameter (curved blade). This process is done to confirm the reliability of the computational study and elucidate the blade's hydrodynamic in detail before assigning a rower force profile.

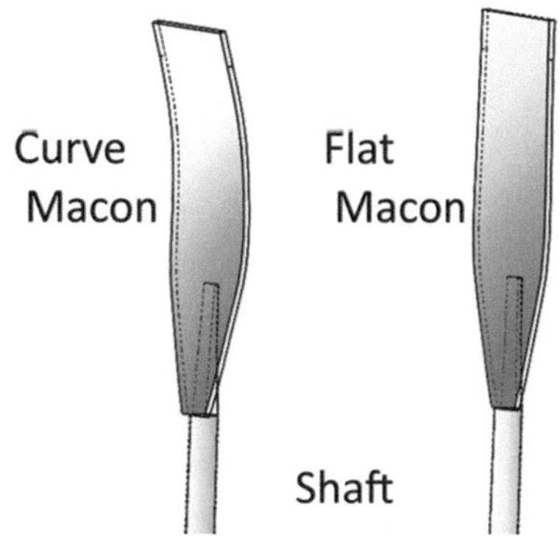

Fig. 6.11 Curve and flat Macon blade design

6.9.1 Result of Quasi-static Investigation of the Blade

In the quasi-static simulation, nine positions of difference angle of attack were selected, ranging from 0° to 180°. The positions were investigated computationally based on the quasi-static approach. Flow velocity was set at 0.75 m/s for the quarter size, and the velocity was increased to 5 m/s, the same as the average value of real condition as studied by Liu et al. [12]. Lift and drag coefficient comparisons were done between quarter-scale and full-scale blades, as in Fig. 6.12. However, maintaining the same graph sine and cosine for the lift and drag coefficient, quarter and full-scale Macon confirms that the blade's size affects the force generated. The graph shows that increasing the scale of the blade and changing the water flow accordingly reduce the coefficient. However, as the paper is interested in studying the full scale of the Macon blade, most of the results will be focused on this scale, and others will be compared.

In quasi-static conditions, the blade was held at a certain angle of attack. Meanwhile, water flows at 5 m/s and then hits the blade. Figure 6.13 illustrates water flow with a straight streamline before it separates as it passes the blade. The Separation of the flow increases at an increasing angle, which is the maximum separation at a 90-degree angle. The Macon blade was placed perpendicular to the fluid motion at this position and blocked some fluid flow.

The hydrodynamic force acting on the blade is the function of fluid velocity, blade projection area, water density, and force coefficients. Maintaining all the parameters and changing the projection area by manipulating the hitting angle, sin, and cosine graphs of lift and drag force were obtained. According to Fig. 6.14, the drag force acting on the blade is higher than the lift force. This means that most of the hydrodynamic force working on the blade is converted to drag force, and some are converted to force lift force.

The drag force generated is a symmetrical bell-type pattern where the magnitude of the force is increasing with increasing angle. When the angle reaches the maximum value, 1855 N at a 90-degree angle, the value decreases to 0 N at 180°. However, for the lift force, there are two maximum lifts, 836 and − 832 N, at the opposite

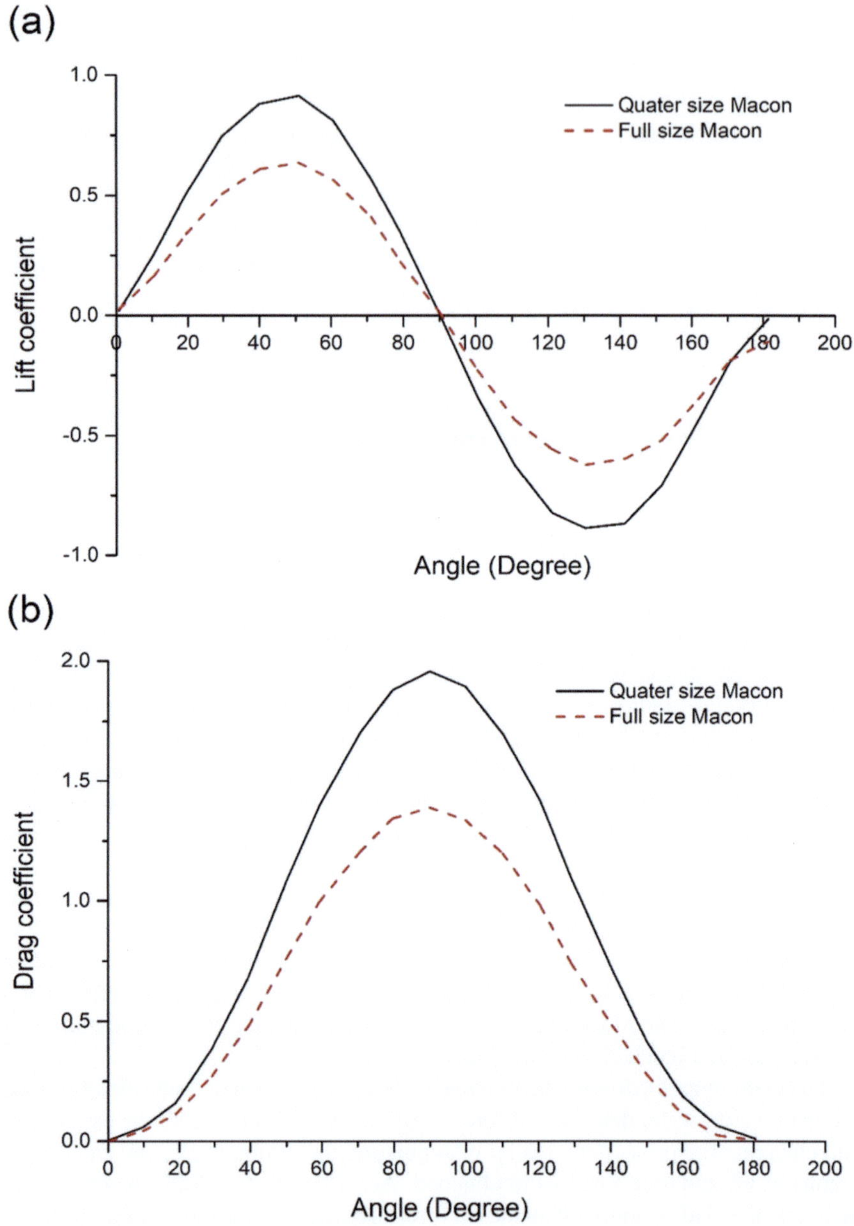

Fig. 6.12 **a** Lift coefficient for the quarter and full-size model. **b** Drag coefficient for the quarter and full-size model

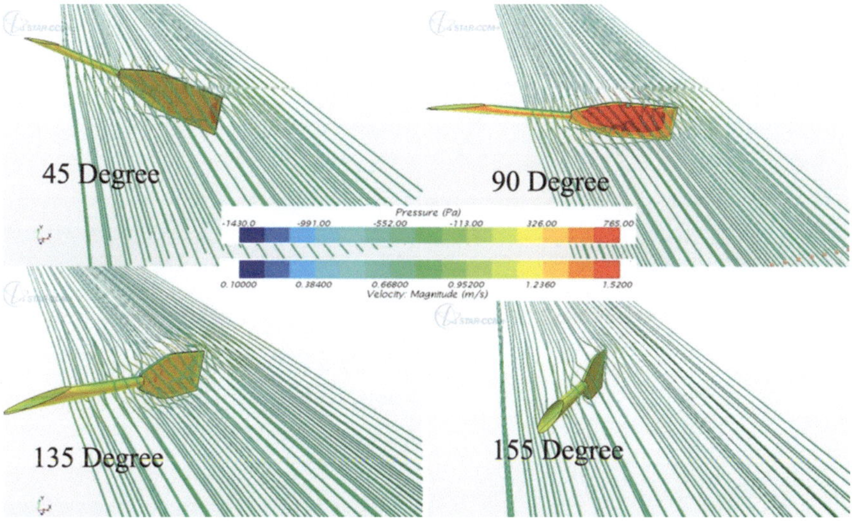

Fig. 6.13 Water streamline and pressure generated on Macon oar's blade at four quasi-static positions

direction, which denote negative and positive values. This happened due to the lift force's changing direction as it passed 9° of attacking angle. The value will be the reference point to the computational study in the next simulation.

6.9.2 Result of Dynamic Blade Rotational Investigation

While rotating, the front surface of the blade hits the water and generates hydrodynamic pressure, as illustrated in Fig. 6.15. Rowing hydrodynamics can be divided into two components based on the blade axis. The first force is normal hydrodynamic stress, which acts perpendicular. The stress acting on the blade moves the boat forward. The force at the blade's front surface is positive, denoted by red colour, and is higher than the surrounding fluid pressure.

Referring to Fig. 6.16, the negative force acts on the blade at the early stroke as the boat moves faster than the blade velocity based on the normal and shear. Still, as the force increases, the blade rotates with higher angular velocity, thus resulting in higher relative velocity, generating positive and more extensive normal stress on the blade surface. Note that the positive sign indicates when the boat moves forward by thrust force. After a specific time, the normal force decreases until the finishing phase. Shear stress, the second component of hydrodynamic force, behaves differently. With a similar pattern to the sine graph but a much lower force magnitude, shear stress contributes significantly to the blade hydrodynamic.

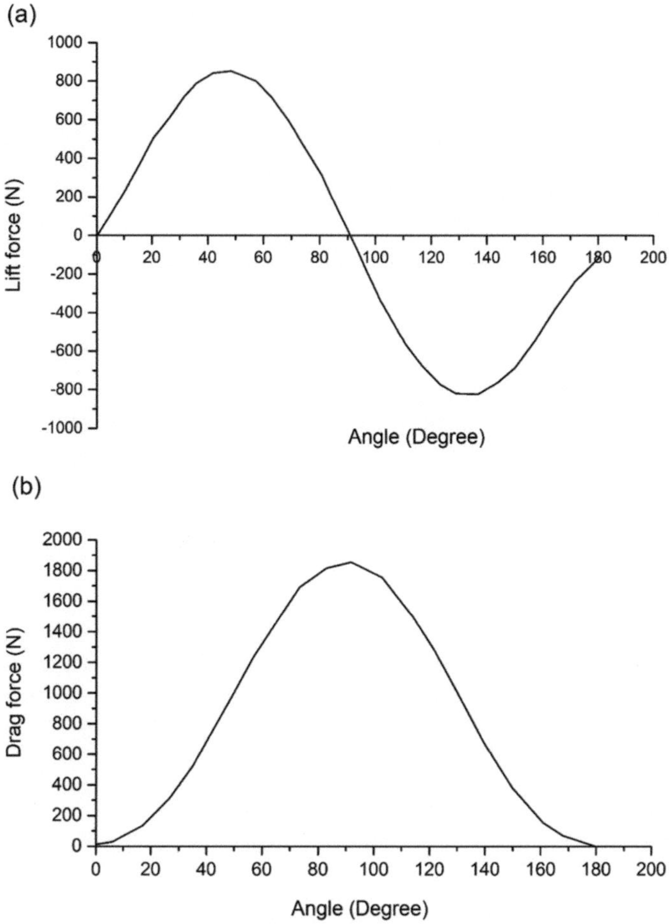

Fig. 6.14 **a** Lift force acting on the full-scale Macon blade. **b** Drag force acting on the full-scale Macon blade

To clarify the effect of the rower stroke on the performance, normal and shear force obtained from the simulation was converted into lift and drag force. The dynamic drag force generated was approximately the same graph pattern as the quasi-static result, as shown in Fig. 6.17. The lift force was shown in the overturn asymmetry pattern of the sine graph, in which the early stroke produced a higher force. However, when the blade passed a certain angle, the value was reduced to negative for the force profile but with a small lift force increment captured toward the finish.

A remarkable difference is observed between the lift and drag force of these three force profiles. During the drive, the blade was moved from catch to finish. Hence, to relate the relation between the rower force profile and the hydrodynamic of the blade, the rowing stroke is divided into the first half and second half of the stroke. The first half of the stroke is the phase where the blade moves from catch to the 90-degree

Fig. 6.15 Hydrodynamic pressure around the blade during stroke

attacking angle and the second half where the blade moves from 90° to the finish. Since the study's interest in the effect of rower force profile on fluid dynamics during the drive, the first half of the stroke and the second half were applied and further discussed.

During the first half of the stroke, negative force acts on the front blade as the water flows faster than the angular velocity of the blade rotation. The angular velocity increases as the applied force increases until it hits the highest point. The highest negative force for drag and lift happens to the first force profile between those three. The negative force acting on the blade surface seems similar for the second and third force profiles. The peak force was achieved early for Force Profile 1, which is 851 N at 31% of the stroke.

Meanwhile, the Second and Third Force Profiles delay the time to achieve the peak by 0.11 and 0.11 s, which is 42 and 39% of the stroke. By looking at the peak character of the lift and drag, the peak point is identified for the second force profile. The horizon peak line was determined as the applied force maintained for a certain period at maximum intensity. Significant differences between the second and third force profiles can be observed since the first force profile pattern is considerably divergent from both.

Meanwhile, the second half of the stroke happened between the peak of the applied load and the finish, showing separate patterns of reducing force. In this aspect, the starting point for the second half of the stroke begins just after the maximum angular velocity. As mentioned, the maximum angular velocity occurs at 0.21, 0.28 and 0.27 s for the first, second, and third force profiles. At the same time, the maximum

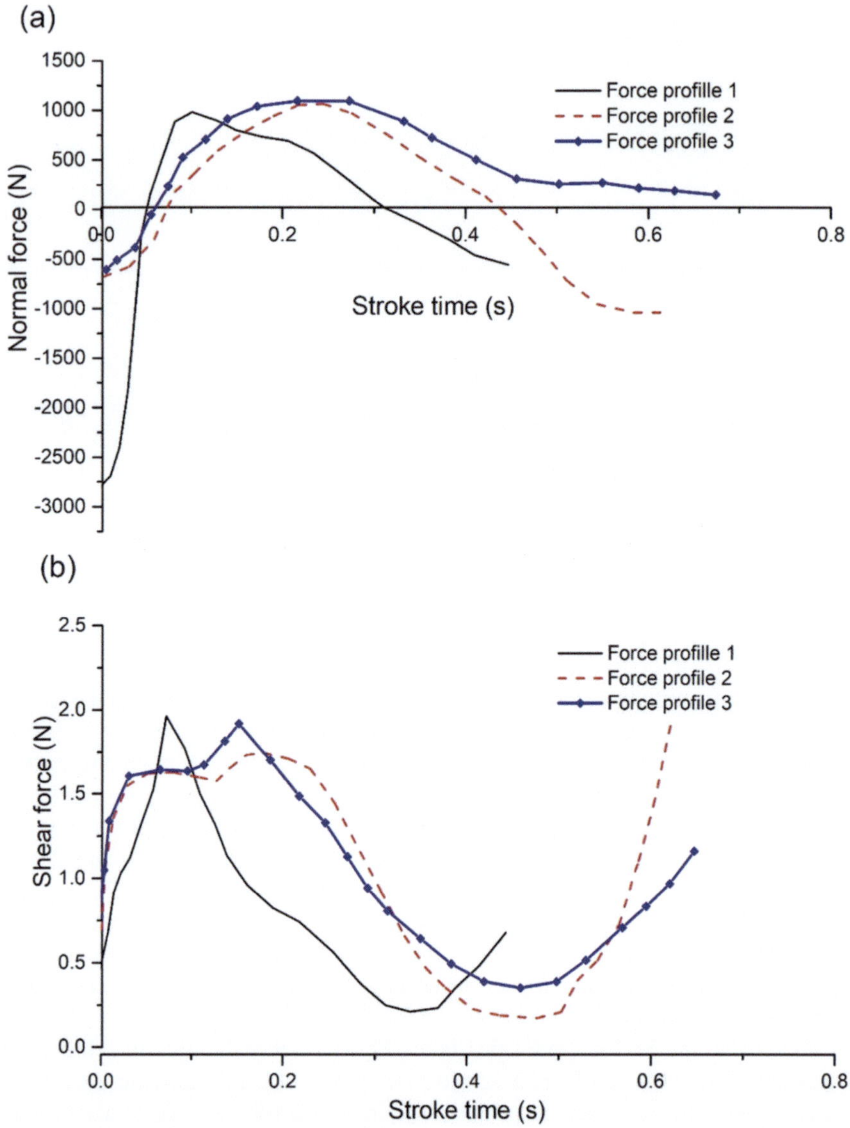

Fig. 6.16 **a** Normal force acting on the blade due to the rower force profile during stroke. **b** Shear force acting on the blade due to the rower force profile during stroke

drag force occurs, whereas the maximum lift occurs at the early time, which ranges between 0.09 and 0.2 s, which was 21–32% of the drive phase. The magnitude of drag and lift is reduced once blade velocity passes the maximum value and rotates with a lower angular speed.

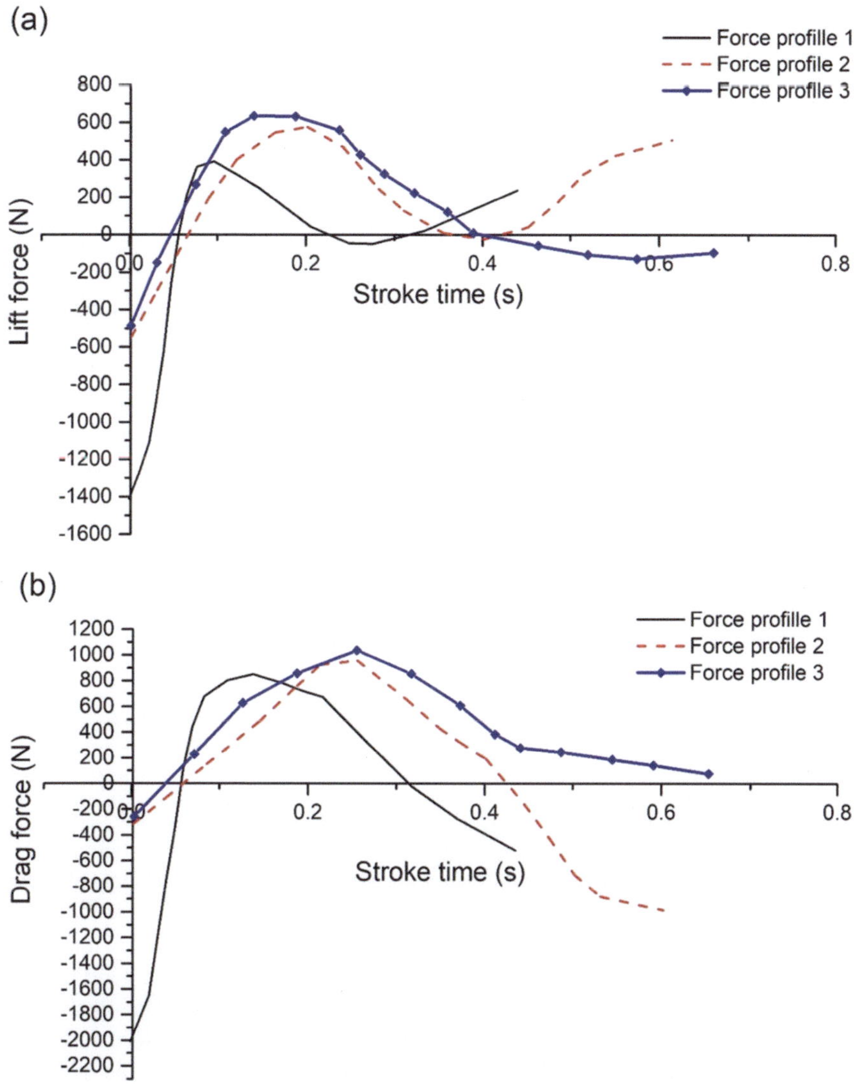

Fig. 6.17 **a** Lift force acting on the blade due to the rower force profile during stroke. **b** Drag force acting on the blade due to the rower force profile during stroke

Generally, the angular velocity of the oar affects the hydrodynamic force. A higher rotation angle generated higher lift and drag. The first and second force profiles continuously decline in drag value until they reach the negative value at the difference points. However, as the third profile, the drag force shows a positive value decline until the end. On the other hand, lift force shows the periodic reciprocate pattern, continuously declining value until the reverted point. The graph goes up constantly until the finish point is reached. Diverse from the first two graphs. The lift force value for the third graph shows a reduction until it becomes a negative value, which is maintained until the finish.

It is important to note that the sweep oar's maximum force acting at different angles compared to the quasi-static study. In a quasi-static study, the maximum lift force acted at the 45-degree angle, whereas drag acts at the 90-degree angle, as shown in Fig. 6.18. However, in the dynamic study, the catch happens at different positions depending on the rower force profile. The catch angle of the Force Profile 1 happens at 54.7°. Meanwhile, the catch angle of Force Profile 2 and 3, respectively, is 30°. Consequently, when the average angle was 56°, the drag force acting at the angle was 66°. Blade position affects the hydrodynamic at a different angle because, in the dynamic study, two velocities, blade rotation and boat translation speed, were considered, whereas in quasi-static, only boat translation was considered.

In the competition, the curved oar blade version was preferred as it could grab more water and push the boat forward. The curved blade water is expected to perform better than the flat version. According to Fig. 6.19, the curve blade's lift and drag force magnitude is higher than the flat blade. Although they force increment because the curve was slight, the values are significant, as the winner is only a few seconds different. Based on Fig. 6.19, Constant elevation was observed in most of the Force Profile's first half of the stroke. For the second half of the stroke, the value of the there is force reduction until the fishing is similar to the previous result.

6.10 Discussion

The simulation carried out in the study is to study the relationship between the rower force profile and the hydrodynamic of the blade. To achieve the objective, the hydrodynamics of the blade were focused on using three models of the blade (rectangular, quarter-scale Macon, and full-scale Macon). This process is important to ensure the computational fluid dynamic is reliable and can produce good results.

Three patterns of the rower force profile were studied. The numerical result showed that the rower's force profile is important in the rowing performance. The result proves that the force profile pattern had a significant effect on the hydrodynamics of the blade, which shows that increasing the rower force will automatically increase the hydrodynamic force. The discussion focused on the performance of the lift and drag force. In the first half of the stroke, the best technique to be applied is to steadily increase the angular speed instead of rapidly growing the force. Newton's first law of motion mentions resistance to movement or inertia, which means that some rapidly increased force used to accelerate the boat is used to overcome the boat and the rower's initial moving state. During that moment, the blade rotates with the minimum angle, leading to higher drag and negative force due to the hydrodynamic resistance of boat velocity. Once the blade rotates with the highest angular velocity,

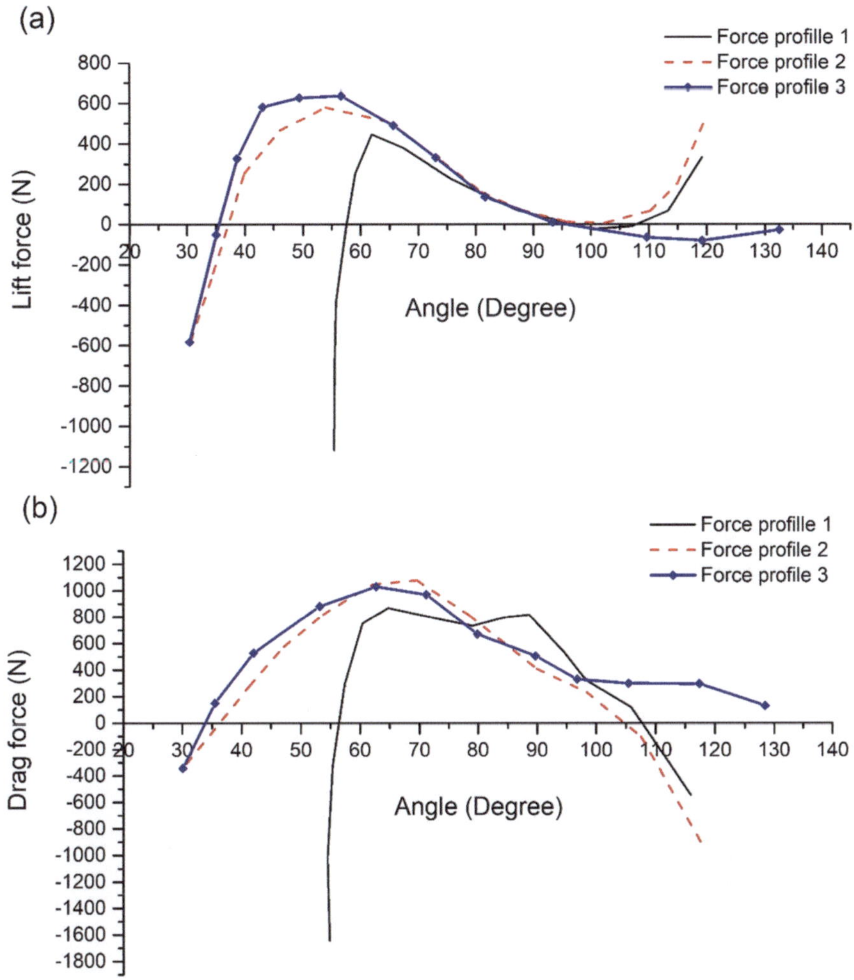

Fig. 6.18 **a** Lift force acting on the blade due to the rower force profile during stroke based on the swept angle. **b** Drag force acting on the blade due to the rower force profile during stroke based on the swept angle

it is an excellent strategy to maintain that condition for a certain period, as it is the applicable action to generate higher work based on hydrodynamic force.

The second half of the stroke is important to complete and provides extra time for hydrodynamic force. Similar to the stroke's first half, higher angular velocity by increasing applied force leads to better hydrodynamic performance. However, the generated force was reduced as the blade moved toward the finish. Numerically, the hydrodynamic result for the three force profiles is noticeably different. Three

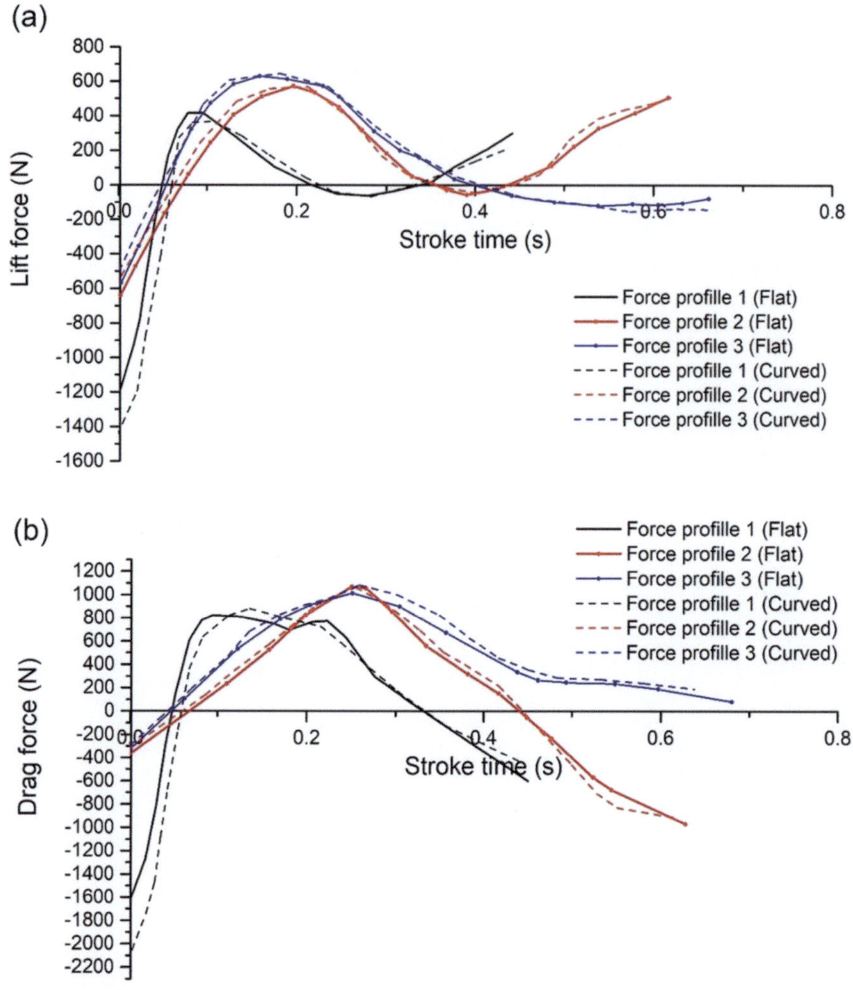

Fig. 6.19 Hydrodynamic forces comparison between flat and curved blade **a** Lift force acting on the blade due to the rower force profile during stroke. **b** Drag force acting on the blade due to the rower force profile during stroke

approaches are suggested for this segment to maximize the hydrodynamic force, especially lift force: (1) reduce the load gradually and gently, (2) Maintain the load at maximum to produce higher angular force, and (3) extend the stroke time to achieve maximum sweep angle.

The maximum lift force for the dynamic study happens at the average angle of 56°, 11° bigger than the quasi-static study at 45°. This situation occurs due to the combination of rotation blade and boat translation speed. Regarding Force Profile 1, it is important to notice that the catch's attacking angle is 54°. By default, the highest force is achieved at almost the same catch time. Meanwhile, the maximum drag force is at an average angle of 66°, 24° smaller than the quasi static study. At this point, the dissimilarity happens due to blade design features. Thus, as the attacking angle of the highest lift and drag is almost the same for the difference force profile, the rower can improve the performance by keeping the highest force a few moments before and after the 56-degree angle.

References

1. Barrett, R.S., Manning, J.M.: Relationships between rigging set-up, anthropometry, physical capacity, rowing kinematics and rowing performance. Sports Biomech./Int. Soc. Biomech. Sports **3**(2), 221–235 (2004)
2. Doyle, M.M., Lyttle, A., Elliott, B.: Comparison of force-related performance indicators between heavyweight and lightweight rowers. Sports Biomech. **9**(3), 178–192 (2010)
3. Baudouin, A., Hawkins, D.: Investigation of biomechanical factors affecting rowing performance. J. Biomech. **37**(7), 969–976 (2004)
4. Pilgeram, K.C., Delwiche, M.J.: Device for on-the-water measurement of rowing output. Sports Eng. **9**(3), 165–174 (2006)
5. Kleshnev, V.: Propulsive efficiency of rowing. In: Proceedings of the XVII International Symposium on Biomechanics in Sports (1999)
6. Kleshnev, V.: Stroke rate vs. distance in rowing during the Sydney Olympics. Aust. Inst. Sport Aust. Rowing **24**(2), 18–22 (2001)
7. Kleshnev, V.: Vertical handle force. Rowing Biomech. Newslett. **13**(143) (2013)
8. Schwanitz, P.: Applying biomechanics to improve rowing performance. FISA Coach **2**(3), 1–7 (1991)
9. Sliasas, A., Tullis, S.: The dynamic flow behaviour of an oar blade in motion using a hydrodynamics-based shell-velocity-coupled model of a rowing stroke. Proc. Inst. Mech. Eng. Part P: J. Sports Eng. Technol. **224**(1), 9–24 (2010)
10. Leroyer, A., et al.: Experimental and numerical investigations of the flow around an oar blade. J. Mar. Sci. Technol. **13**(1), 1–15 (2008)
11. Kleshnev, V.: Oar inertia forces. Rowing Biomech. Newslett. **170** (2015)
12. Liu, Y., et al.: Increased foot-stretcher height improves rowing performance: evidence from biomechanical perspectives on water. Sports Biomech. 1–12 (2018)

Chapter 7
Design and Setup of Rowing Water Tank for Data Collection

7.1 Introduction to Dynamic Rowing Simulator

The main factors affecting the rowing performance are the rower's biomechanics and the blade's hydrodynamic force [1–6]. The biomechanics of the rower can be classified as the kinetics and kinematics of a rower [7–9]. There are disconnected studies between the rower biomechanics and the blade's hydrodynamics, where the boat speed prediction is not comprehensive. This is due to the complexity of the rowing system, which leads to the difficulty in analyzing the rower's biomechanics and blade hydrodynamics. Moreover, most of the investigations analyze the biomechanical aspect directly with the boat speed by neglecting the hydrodynamic effect on the oar blade or the other way around [3, 10].

Since a limited number of studies focus on the coupling mechanism between the biomechanics of rowing and the hydrodynamics of the blade, the current study was carried out. The study's objective is to evaluate the coupling mechanism between rower biomechanics in the aspect of rower's kinematic and handle force to hydrodynamic of the blade of the drive phase, which is the first research objective, as mentioned in Chap. 1.

The special dynamic rowing simulator was used for this purpose. The dynamic rowing simulator was equipped with equipment and software designed to simulate the experience of rowing on water in an indoor setting while collecting the data for research and performance improvement. The device aims to provide a more realistic rowing experience by replicating the movement of a rowboat on water. It could achieve this through various mechanisms, such as incorporating a moving seat or sliding rail system, allowing the user to engage more muscles and experience a more excellent range of motion during the rowing stroke.

By attaching the sensor to the dynamic rowing simulator, important data can be collected to track and enhance an athlete's performance. Gathering and interpreting the data can provide valuable insights into an individual's strengths, weaknesses, and progress. Some common data collected for on-water rowing training

© The Author(s), under exclusive license to Springer Nature Singapore Pte Ltd. 2025
A. A. Mohd Yusof et al., *Rowing Biomechanics and Hydrodynamics*,
Series in BioEngineering, https://doi.org/10.1007/978-981-97-7271-1_7

include the number of strokes a rower takes per minute. Monitoring stroke rate helps athletes maintain an optimal pace and rhythm during rowing sessions. In addition, the collected force curves show how force is applied during each stroke, allowing coaches to identify areas for improvement in power application and timing. Besides, measuring the amount of power a rower generates during each stroke can be used to assess performance and progress.

7.2 Design and Schematic Diagram

Following is a schematic diagram representing the components and connections within a system to measure the forces and speed of the dynamic rowing simulator. The system involves water piping and electronic sensing systems, as in Fig. 7.1.

As in the schematic, five elements are included for the sensor system, as in Table 7.1.

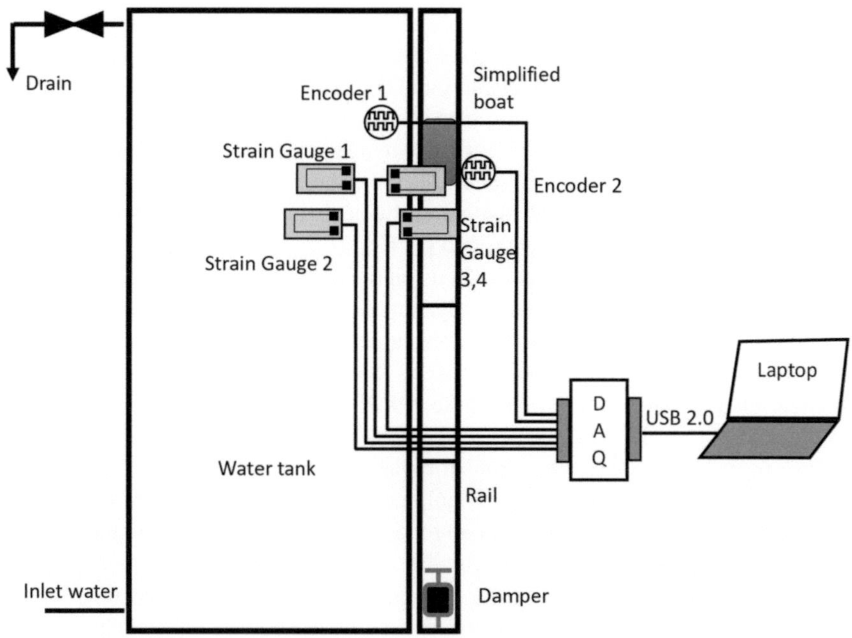

Fig. 7.1 Schematic diagram for dynamic rowing simulator

Table 7.1 Important elements of the sensing system

System	Description
Inboard sensing	The primary components of the system that measure and detect the desired physical quantities of rower handle force using two strain gauges attached close to the handle
Outboard sensing	Blade force corresponding to the rower handle force is measured using two strain gauges attached to the outboard shaft
Oar rotational Speed	Oar movement during stroke was measured using an encoder located on the Oarlock
Boat speed	Since the dynamic rowing simulator has been replaced with a simplified version of the boat, the speed during the stoke is captured using an encoder that is attached to the roller
Data acquisition systems	This system enables the sensors to transmit the reading data to a central monitoring system and present the measured data in a human-readable format, such as an LCD

7.3 Design Calculation

The reliability of the simulator structure refers to its ability to perform its intended function without failure or excessive deformation. During data collection or testing, pool structure and rail experience high dynamic load, which is important to be assessed earlier during the designing process. Assessing the reliability of a structure involves evaluating its design, construction, and performance characteristics to determine the likelihood of failure or degradation related to its working process.

Based on the preferred location, the dynamic rowing simulator was designed from several modules and joined using a fastener. It is expected that the design will follow the above-ground water tank. Generally, the dimension of the water was 15 m × 15 m × 1.2 m. Thus, the main structure will stand the higher applied load.

The maximum hydrostatic pressure on the water acts on the wall of the water tank.

$$\boldsymbol{P_h = \rho g h_c} \tag{7.1}$$

The equation used to estimate above-ground water tank type.

Where

Density of water, $\rho = 1000\,\text{kg/m}^3$.

Gravity acceleration, $g = 9.81\,\text{m/s}^2$.

Height of water, h_c.

7.3.1 Calculation for Hydrostatic Force Generated

$$\text{Ph} = 1000\,\text{kg/m}^3 \times 9.81\,\text{m/s}^2 \times 1.2\,\text{m} = 0.118\,\text{MPa}$$

Maximum *pressure acted at the bottom of the tank and wall.*

Force acting on the side wall.

Assuming hydrostatic pressure acts in a linear gradient (Max on the bottom and zero on the surface). Thus, the average pressure can be obtained by integration of the linear gradient or 2/3 of the height.

Average hydrostatic pressure,

$$\text{P Avg} = 2\text{Ph}/3 = 2(0.118\,\text{MPa})/3 = 0.079\,\text{MPa} \tag{7.2}$$

where Area of the Water Tank Wall,

$$\text{A} = 1.2\,\text{m} \times 15.0\,\text{m} = 18\,\text{m}^2$$

Average hydrostatic force,

$$\begin{aligned} \text{F Avg} &= \text{Pavg} \times \text{A} = 0.079\,\text{MPa} \times 18\,\text{m}^2 \\ &= 1.42\,\text{MN} \text{ @ } 1420\,\text{kN} \end{aligned} \tag{7.3}$$

7.3.2 Stability of the Frame Structure

The stability of the frame involves Second-moment inertia, in which the water tank frame structure was made from rectangular hollow 3-inch (0.076 m) × 2-inch (0.051 m) with a thickness of 1.6 mm.

$$I_x = \frac{bh^3}{12} = \frac{0.051\,\text{m} \times \left(0.076^3\right)}{12} = 1.87 \times 10^{-5}\,\text{m}^4 \tag{7.4}$$

Half of the frame thickness, $y = 0.076\,\text{m}/2 = 0.038\,\text{m}$.

Where there were eight vertical frames for each wall, and d refers to the average location for force act 1/3.

Moment act on the frame, $\text{M1} = \text{F} \times \text{d} = (1420\,\text{kN}/8) \times (1/3 \times 1.2\,\text{m}) = 71\,\text{N m}$.

7.4 Assembly Drawing of Dynamic Rowing Simulator

Figure 7.2 shows a detailed view of the rowing dynamic simulator. It provides a detailed view of how individual parts come together and interact to form a working simulator. These assembly drawings are commonly used in engineering, manufacturing, and design fields to communicate how different parts fit and work together within a larger structure. In total, it has 16 main components (RS-01 to RS-16). The biggest part of the dynamic simulator is the water tank (the combination of RS-02 and RS-03). The rail (RS-01) was then attached to the side. The design is flexible, and the water tank and rail can be extended or shortened depending on the requirement.

The most complicated part is the Boat frame (RS-04). Almost all the parts are attached to the boat frame to ensure that important data can be captured. Table 7.2 shows a list of the components of the dynamic rowing simulator.

Three important parts are specified in Figs. 7.3, 7.4, and 7.5 for the component detail. Six pillars were used to build the frame on both the left and right sides. The pillars ensured the railway was strong enough to withstand the water tank's simplified boat movement and vibration. Six horizontal frames were used to connect both the left and right side frames. At the same time, the horizontal frames acted to cope with the force generated during the test. The frames were made from square hollow steel with a size of 2-inch × 3-inch and a thickness of 1.6 mm.

M10 bolt and nut used to fasten all the frames. The bolt threads were checked to ensure they were properly engaged in the nut. A general rule of thumb applied was that the threads' engaged length should equal the bolt diameter.

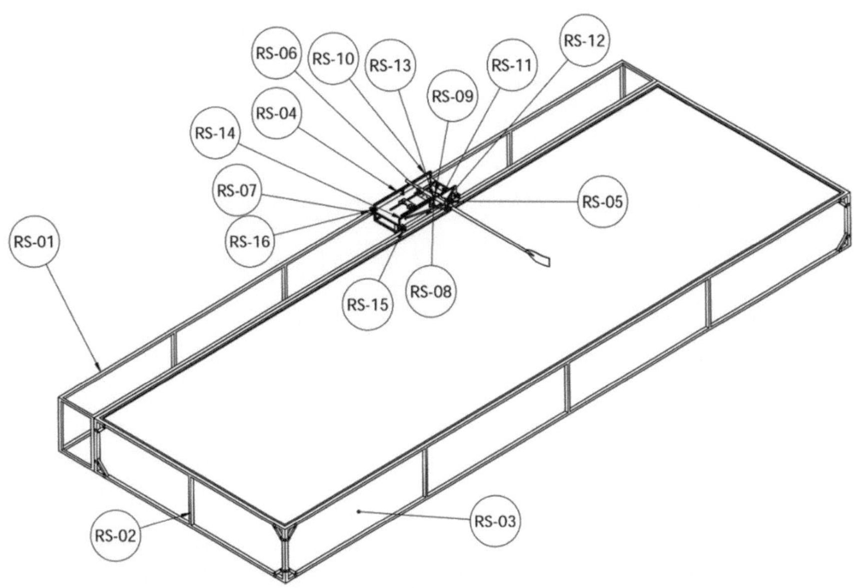

Fig. 7.2 Assembly drawing of the dynamic rowing simulator

Table 7.2 List of the drawing and detail

Drawing No.	Quantity	Description
RS-01	1	Rowing rail
RS-02	1	Pool frame
RS-03	1	Pool canvas
RS-04	1	Boat frame
RS-05	1	Potentiometer holder
RS-06	1	Footrest base
RS-07	1	Seat base
RS-08	2	Stopper ring
RS-09	1	Potentiometer rod
RS-10	1	Velocity sensor holder
RS-11	1	Bend rod
RS-12	1	Locker rotational sensor
RS-13	1	Protentional ring
RS-14	1	Ringing base
RS-15	2	Roller holder right
RS-16	2	Roller holder left

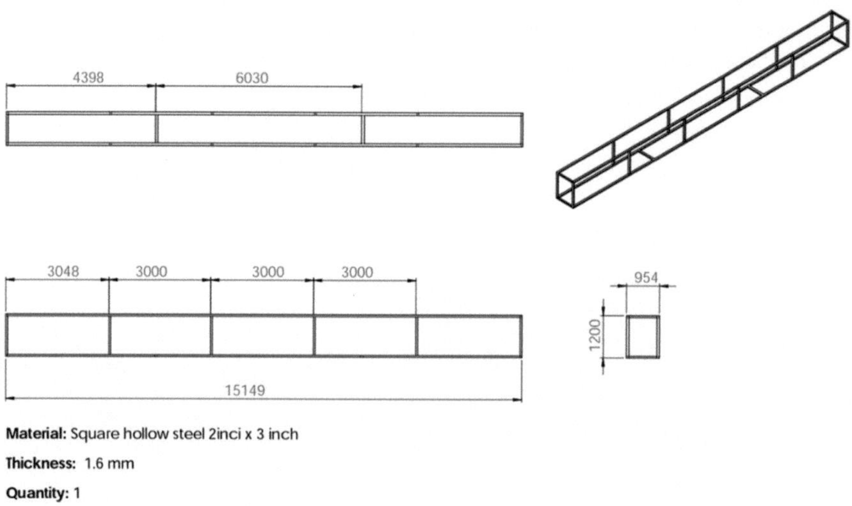

Fig. 7.3 Rail design for the dynamic rowing simulator

The frame of the water tank was built as in Fig. 7.4. The structure was divided into two sections: section A, the main frame, and section B, the supporting frame. For section A, since it was the main structure of the water tank, the frame was made from rectangular hollow steel, 3-inch × 2-inch. At the same time, section B was

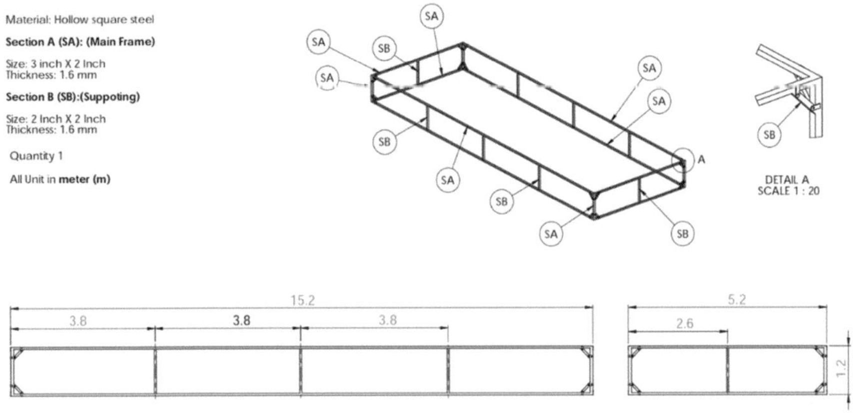

Fig. 7.4 Water tank frame structure

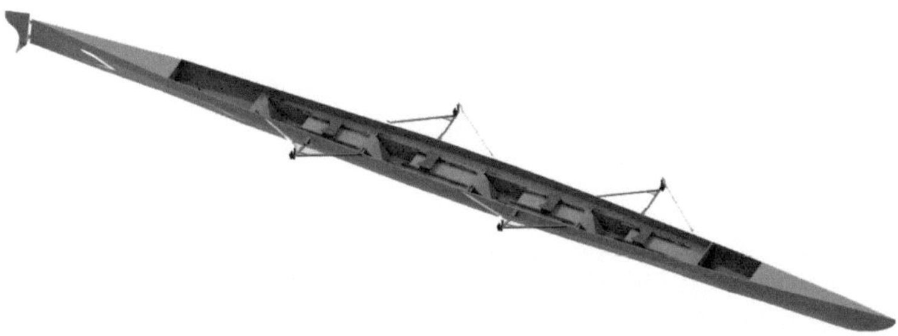

Fig. 7.5 The original boat model used as a reference for the simplified rowing boat

made from rectangular hollow steel of 2-inch × 2-inch. Besides that, to increase the frame's rigidity and prevent undesirable deformation, corner brackets were installed on each corner using a 45-degree angle steel plate.

Figure 7.5 shows the original boat borrowed from PERDAMA, Malaysia. To simplify the original rowing boat into a simplified simulator, the size of the boat was reduced, and the important features were maintained. The single-seat design was then included with minimal controls. Focus on fundamental rowing mechanics, limiting the complexity of water simulations and environmental factors. This scaled-down version as a simulator ensures a user-friendly, uncomplicated rowing simulation, making it accessible for users of all skill levels while retaining the essence of the engineering aspect of the rower biomechanics and hydrodynamic of the rowing, especially for the blade during stroke.

Figure 7.6 shows the frame for the simplified boat. The frame dimension follows the real boat measurement with minor modifications to ensure it fits the rail and water tank. The frame was fabricated using square hollow steel (1 inch × 1 inch). The length of the frame was 1500 mm, and the width was 680 mm. The size fits for one sweep oar at the side. The thick board will be placed at the side as a base of

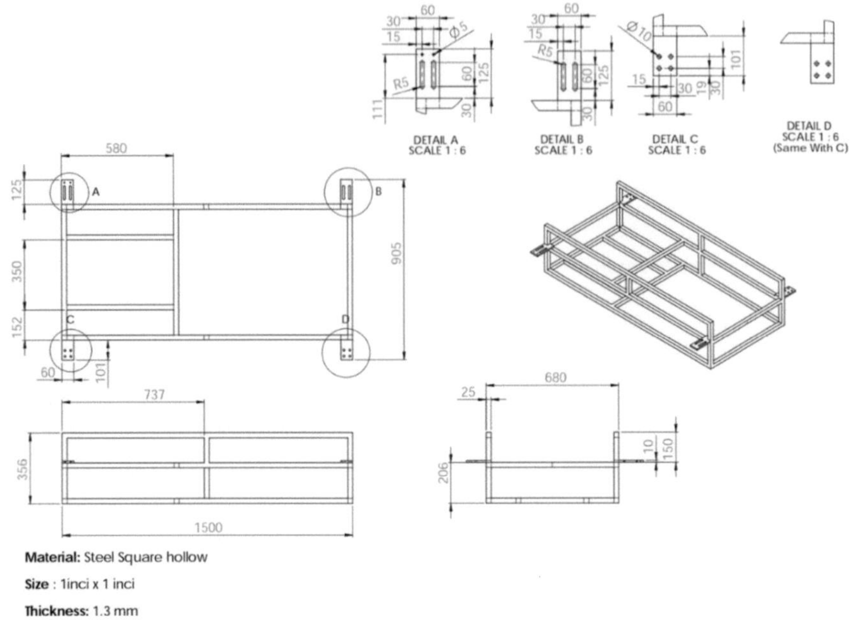

Fig. 7.6 Simplified boat frame structure

the rigger. Rollers were included in the frame design since the simplified boat was placed on the rail. As shown in Fig. 7.6, there are four detailed drawings (Detail A, B, C, and D). Detail A and B are roller bases with long holes for adjustable purposes to ensure the moving simplified boat properly attaches to the rail.

7.5 Blade Design for Dynamic Rowing Simulator

The design of oar blades plays a crucial role in rowing, affecting both the efficiency of rowing strokes and the overall performance of the rower or crew. Oar blades, also known as oar blades or oarlocks, are the flat, spoon-shaped attachments at the end of the oars that are submerged in the water during rowing. These blades are designed to maximize propulsion while minimizing resistance, contributing to efficient and effective rowing strokes.

Ultimately, the design of oar blades aims to strike a balance between power, efficiency, hydrodynamics, and ease of use. Rowers and coaches carefully consider these factors in selecting oar blades that align with their specific rowing goals, techniques, and water conditions. The right oar blade design can significantly enhance performance, reduce fatigue, and provide a smoother rowing experience.

In this topic, three blade designs were tested to investigate the effectiveness of their design based on the stroke styles. The design of oar blades and the stroke style employed by rowers significantly impact rowing performance and efficiency. The interaction between blade design and stroke style influences power output, boat speed, catch efficiency, and overall technique.

As shown in Fig. 7.7, the Macon blade is an oar blade commonly used in rowing. It has a distinctive shape that sets it apart from other blade designs. The Macon blade design offers good stability in the water. It has a relatively large surface area and a more gradual curve, allowing rowers to maintain a stable connection with the water throughout the stroke. The design might not be ideal for all rowers or competitive situations. More experienced rowers or those seeking specific performance gains might opt for blade designs with different characteristics tailored to their needs.

The cleaver blade has a squared-off or hatchet-like tip, unlike the more traditional rounded tip found in blades like the Macon. This squared-off tip resembles the cleaver knife, which gives the design its name, as shown in Fig. 7.8. The design is particularly effective at generating power during the catch phase of the rowing stroke. The squared-off tip allows for a strong and direct catch, making it well-suited for rowers who emphasize power application right from the start of the stroke. The design of the cleaver blade can help minimize air resistance during the recovery phase. The squared-off tip slices through the air with less drag than other blade shapes, which can be advantageous during high stroke rates.

The fat blade refers to a blade design with a wider and larger surface area than traditional blade designs. A wider blade could offer advantages in increased catch efficiency, generating more power, and potentially being more forgiving in maintaining water contact during the stroke. It is designed to maximize the water contact area and power transfer during the rowing stroke, as in Fig. 7.9. A wider blade might help rowers who emphasize strength and power application.

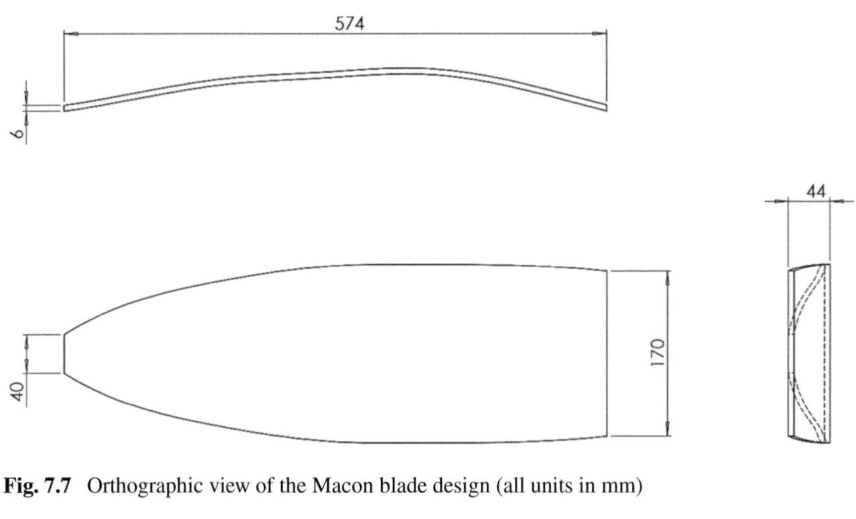

Fig. 7.7 Orthographic view of the Macon blade design (all units in mm)

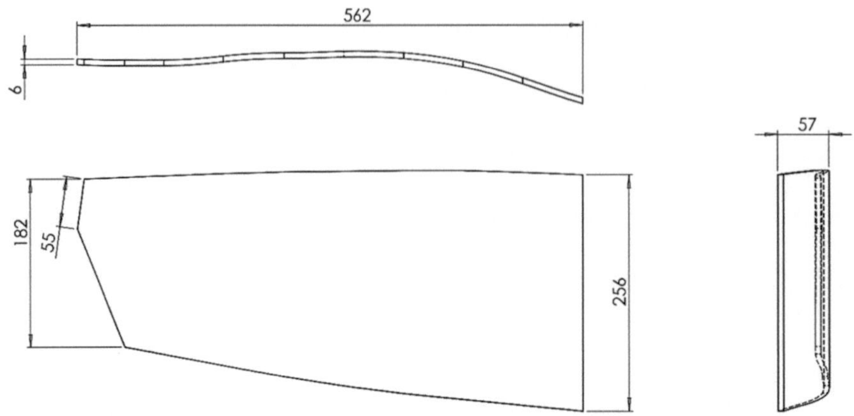

Fig. 7.8 Orthographic view of the cleaver blade design (all units in mm)

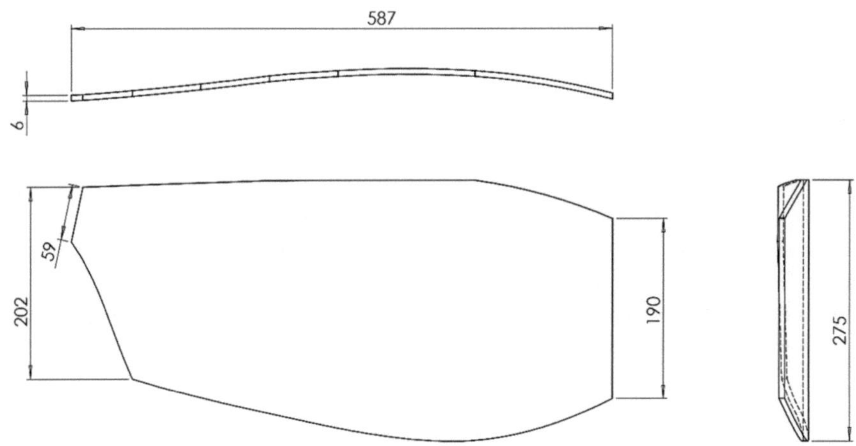

Fig. 7.9 Orthographic view of the fat blade design (all units in mm)

The choice of blade design, whether cleaver or another type, depends on a rower's individual preferences, the type of rowing they are engaged in (sprint, endurance, recreational), and their skill level. Rowers often work with coaches and experts to determine the best blade design that aligns with their goals and rowing style.

7.6 Dynamic Rowing Simulator Troubleshooting During Testing and Data Collection

Troubleshooting was carried out during the simulator testing to ensure the reliability and repeatability of the obtained data. The process is important in considering the effect of wear and vibration on the captured data and making suggestions for improvement. The process involves diagnosing, identifying, and resolving the related issues and problems in the dynamic rowing simulator. The main objective of this section is to restore the product to its normal functioning condition.

7.6.1 *Bending of the Tank Frame Structure Due to Stroke Wave*

The problem happens after a few times of the testing. However, in the design, the structure is expected to withstand hydrostatic pressure up to 1000 kN. However, once stroke is introduced, blade movement generates a dynamic wave, which adds additional dynamic pressure to the water tank structure and causes slight bending about 3 cm from the original. The bending issue can be solved by adding additional support on the bending frame from the affected structure to the floor using an L beam.

7.6.2 *Rail Bending Due to Extra Force from Blade Hydrodynamic*

Frame bending also happened on the rail, especially on the frame side, which was far from the water tank. Once the blade was swept, blade force generated up to 230kN and pushed the boat forward. At the same time, the force component acts perpendicular to the boat's direction. The force component also pushes the rail outward and bends the frame structure. The problem can be solved by adding extra support that connects the inward and outward rail pillars. Thus, simultaneously, the support keeps the distance between the two pillars consistent at 0.94 m.

7.6.3 *Processing Software*

This section involves processing software initially, and the project was planned to use LabVIEW software to capture data automatically and process the boat speed and acceleration spontaneously. However, as the sensing system involves many settings, it sometimes causes an error during the experiment. SignalExpress, provided by DAQ hardware supplied, was preferred to avoid this and make the data collection easier.

Detail software
SignalExpress 2015
Support Os: Window 7
Application Bitness: 32-bit
Language: English
File size: 975.1 MB.

7.6.4 Wear on the Encoder Roller

Wear is the standard issue that happens on the moving component. For this simulator, the most critical part that faces the problem is the encoder roller since it closely touches the simplified boat roller to collect the boat speed. This issue becomes critical for extended use as it causes the roller to slip and affect the reading. To solve the problem, an additional thick rubber of 3 mm is placed in between to increase the contact. For maintenance, the rubber will be changed after 50 tests or in 4 months for optimal performance.

7.7 Experimental Approach

Using a dynamic rowing simulator, the study mimicked the rowing biomechanics and the hydrodynamic condition. This approach eliminated the skin drag on the rowing boat, heave and pitch motion, the balancing effect of the boat, and bow surges generated during the boat speed. The simulator's motion was similar to the rowing boat cruise on the water. However, as the boat was controlled, most of the hydrodynamic force of the blade was utilized to propel the boat and increase its efficiency.

The dynamic rowing simulator used in the experiment consisted of three main parts: (1) the simplified rowing boat. (2) Boat rail. (3) Water tank (refer to Fig. 7.10). The simplified rowing boat moved freely along the rail using roller bearings during the test. Replicating the rigging system of the rowing boat, an oar with a Macon blade was attached to the boat using the oar lock, which was placed at the right wall of the boat.

The size of the simplified boat was 1.50 × 0.69 × 0.36 m in length, width, and height, respectively. The simplified rowing boat and oar weighed 35 and 2.5 kg, respectively. The size of the rail was 15.00 × 0.95 × 0.90 m, designed to fit enough to the simplified rowing boat. The water tank size was 15.00 × 5.00 × 0.90 m for length, width, and height, respectively.

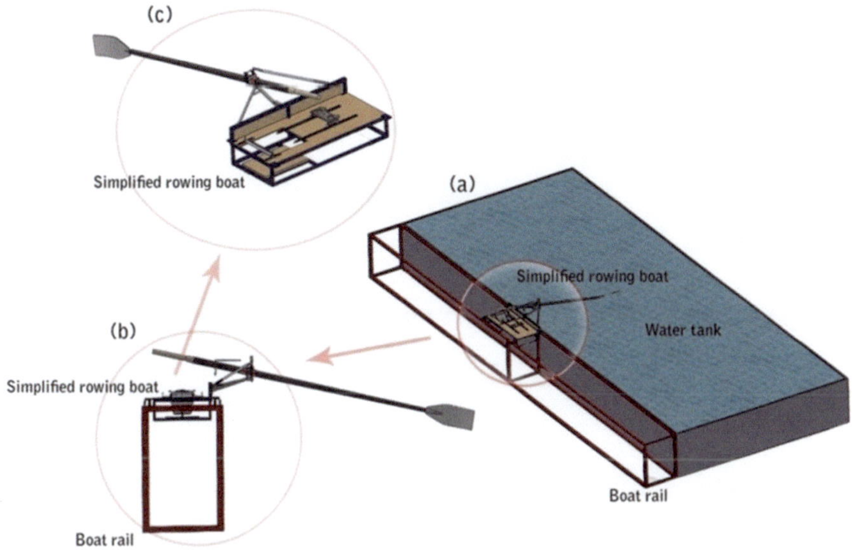

Fig. 7.10 Experimental equipment used in the study; **a** complete setup of rowing equipment, which consisted of the simplified boat, rail, and water tank; **b** model of the rail and how the simplified boat was placed and **c** model of the simplified boat that replicated the real model of the rowing boat

7.8 Description of the Measurement Device

Handle and blade hydrodynamic forces were sensed according to the shaft strain bending [11, 12]. Four foil strain gauges FLA-6-350-17 manufactured by Tokyo Sokki Kenkyujo were used in the experiment. Strain gauges were glued to the shaft using a general-purpose Cyanoacrylate adhesive material. Two strain gauges were located on the outboard shaft, one on the tension and the other on the compression side, 24 cm away from the oar collar [3]. The other two strain gauges were placed inboard at the region experiencing tension and compression. These inboard strain gauges were located 15 cm from the oar collar. The strain gauge installation is shown in Fig. 7.11a and b.

The oar rotational angle and boat velocity were monitored using the rotary encoder DFS60A-S4PC65536 which is manufactured by SICK. The encoder's pulse train signal provided the corresponding angular value, which measures the angular and linear speed. The first encoder was located on top of the gate, as shown in Fig. 7.11a. It was connected to the oar using the long rod attached to the rotating cylinder. The setup allowed the oar to move vertically, rotate to the axis of the oarlock, and turn around to the longitudinal axis. The second encoder was placed on the roller, as shown in Fig. 7.11d. This setup was established to capture the boat's movement during the stroke. All the strain gauges and encoders were connected to a National Instrument data acquisition system with NI 9219 and NI 9401 modules and were slotted into the NI9174 chassis, as shown in Fig. 7.12.

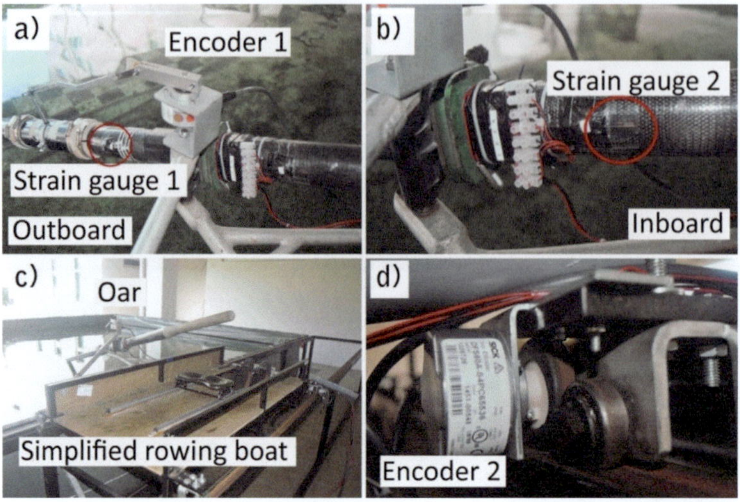

Fig. 7.11 Sensors setup and locations used in the study; **a** sensors placement on the outboard side, **b** sensor placement on the inboard side, **c** simplified rowing boat on the rail, and **d** encoder placement on the simplified boat to read the boat velocity

Fig. 7.12 Sensor setup and connection

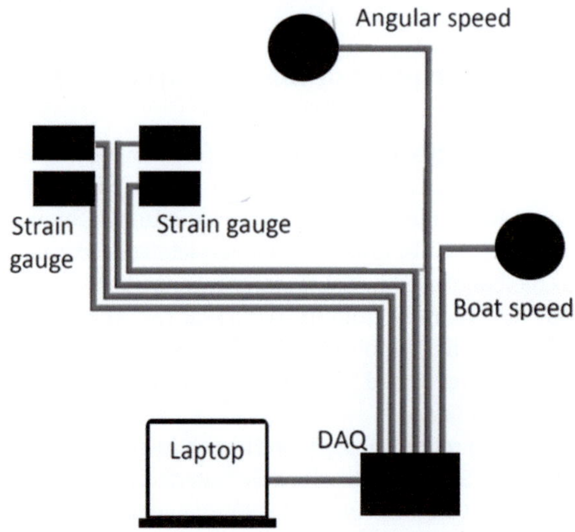

7.9 FEA of the Oar Bending for Strain Gauge Placement Location

Based on the previous study, the strain gauge placement's strain contour was reconfirmed using Finite Element Analysis (FEA). ABAQUS FEA software (ABAQUS Inc., USA) was used to access the strain distribution of the oar due to the bending

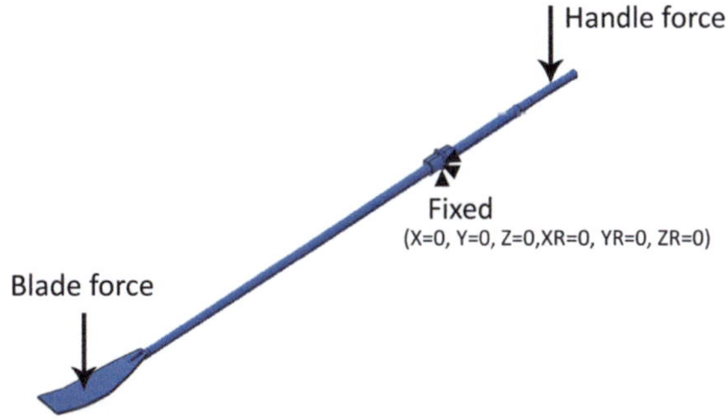

Fig. 7.13 Oar model and boundary condition assigned for the FEA study

moment. The boundary condition of the simulation replicated the oar fixation on the boat, as shown in Fig. 7.13. The oar button was fixed in all degrees of freedom (X, Y, and Z and rotational axes (XR, YR, and ZR)). By fixing the oar button, which represented the part that fastened to the oarlock, the setting allowed the inboard and outboard shafts to bend without excess constraint. The handle force of 600 N and the blade force of 250 N were applied to the handle and blade of the oar.

The oar was made of three materials: (1) Pinewood for the handle, (2) laminated carbon fibre for the shaft, and (3) epoxy resin for the blade. Material properties of 570 kg/m^3, 12.3 GPa, and 0.3, respectively, for density, Young's modulus, and aspect ratio were assigned for pine wood [13]. Material properties of 1.5×10^3 kg/m^3, 29 GPa, and 0.38 were assigned for Epoxy resin [14, 15]. Meanwhile, material properties of 0.18 kg/m^3, 70 GPa, and 0.49 were assigned for laminated carbon fibre [16]. The overall length of the oar model was 3.81, and the average shaft diameter was 0.06 m. The inboard length from the button part to the end of the handle was 1.15 m, and the outboard length from the button part's tip of the blade was 2.66 m.

The FEA study of oar bending was analyzed according to the static general procedure of large deformation structures. The oar model was meshed using a tetra element with the element type C3D10 (A 10-node quadratic tetrahedron). The strain result was selected for post-processing work to ensure strain gauges were placed at a suitable location.

The contour description of the FEA study is shown in Fig. 7.14. Strain distribution has been selected for the post-processing analysis to represent the strain value read by the strain gauge. It reconfirms the strain gauge placement location provided by the previous study. The principle strain was selected for the analysis work to represent the distribution of the normal strain for a specific point at a particular direction on the structure. In general, the strain occurred on the oar due to the bending of the oar shaft as the force was applied to the handle, and the hydrodynamic force acted on the blade.

Due to the applied load, the oar shaft experienced negative and positive strain, representing compression and tension conditions. The maximum experienced strain that occurred on the inboard was 5.37×10^{-4}.

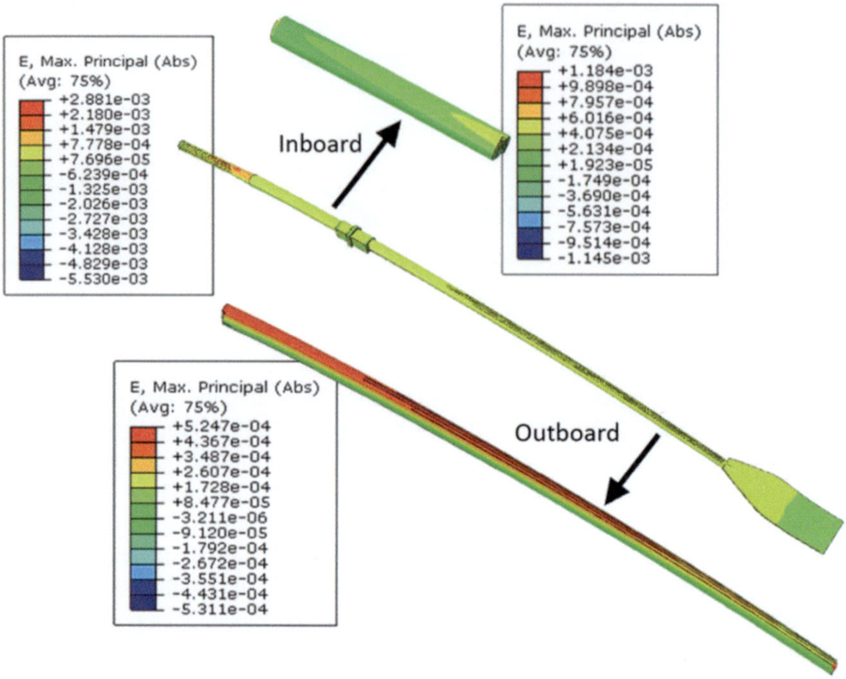

Fig. 7.14 Strain distribution on the oar due to the force applied

Figure 7.15 shows the strain distribution on the oar shaft, measured from the sleeve. The strain magnitude on the inboard shaft was distributed linearly and smoothly, from 0 to 25 cm, illustrated in Fig. 7.15a. The outboard shaft's strain magnitude was randomly distributed slightly, increasing from 0 cm to 32.5 cm, as shown in Fig. 7.15b. The red dot line shows the strain value for the 15 cm and 24 cm, as suggested by the previous study. This strain value is considered relatively higher and is expected to be correctly read by the strain gauge to represent the force applied.

The strain on the inboard shaft and the outboard shaft differed due to the variation in structure rigidity and magnitude of bending moment. The average inboard diameter was 22% higher than the inboard. Thus, it affected the shaft's second-moment area, making the inboard shaft hard to bend. Since the average diameter of the outboard was smaller, the shaft was flexible. Besides, the length of the outboard shaft was double compared to the inboard shaft. Due to this factor, the bending stress on the outboard shaft was higher. The shorter inboard length and higher rigidity structure made the strain distribution steady. The strain distributions of the outboard shaft then varied due to the flexible structure, and a higher bending moment made it less stable.

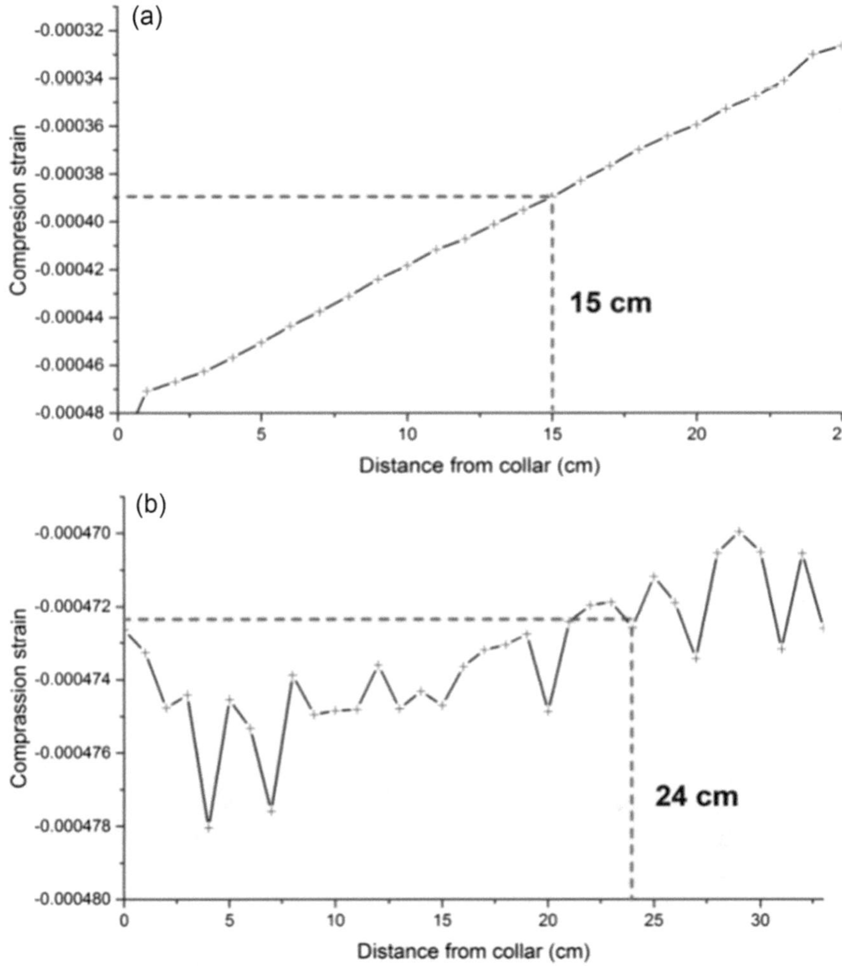

Fig. 7.15 Compression strain happens on the oar shaft due to the force applied to **a** the inboard shaft and **b** the outboard shaft

7.10 Data Acquisition System

Each strain gauge and encoder was connected to National instrument data acquisition modules slotted into NI9174 chassis. NI 9219 was a universal analogue input module with a built-in bridge measurement circuit. Strain gauges were connected directly to the input module port to complete a three-wire-quarter-bridge strain gauge circuit, as in Fig. 7.16 [17]. Using this configuration, strain measurement was measured according to the bridge imbalances indicated by the voltmeter. As the distance between the strain gauge and the bridge circuit was far away, the third wire attached

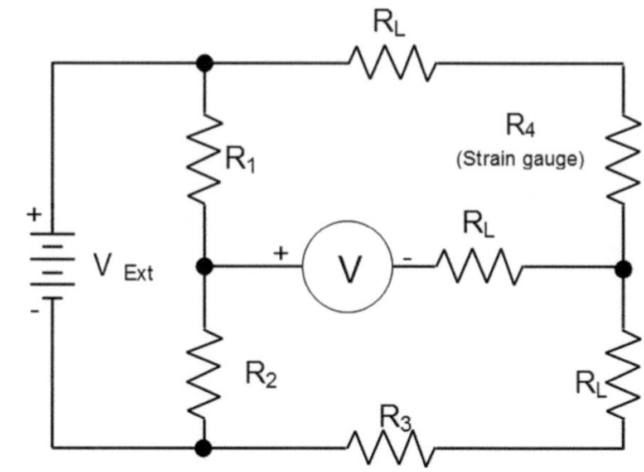

Fig. 7.16 Quarter bridge 1-single active stain gauge

to the voltmeter was connected directly to the strain gauge terminal to reduce the wire resistance effect.

NI signalled express software provided by the manufacturer to transfer the reading data into the computer for visualization and further investigation. The signal from the encoder was converted into the degree of oar rotation by the following equation:

$$\theta = \frac{180}{65536}K \quad (7.5)$$

While 65536 was the number of pulses that occurred for one complete rotation, 180°, and K was the number of pulses read by the encoder during the experiment. The signal from the second encoder was converted into the travelling distance, according to Eq. 7.2.

$$l_b = \frac{2.436 \times 10^6}{65536}K \quad (7.6)$$

Meanwhile, 2.436×10^6 was the ratio of linear distance for one complete rotation. This value was related to the radius of the roller attached to the encoder.

7.11 Calibration of the Strain Gauge

The study is interested in obtaining the force magnitude acting on the oar handle and the corresponding blade hydrodynamic force to investigate the oar's coupling mechanism. Strain gauges were calibrated using the dynamic load [12, 18]. The inboard side of the oar was properly fixed by gripping the oarlock and oar shaft [19]. The oar calibration setup using a mechanical testing machine (Instron 8874 with 25 kN load cell) is shown in Fig. 7.17. The machine pushed the blade part at 50 mm

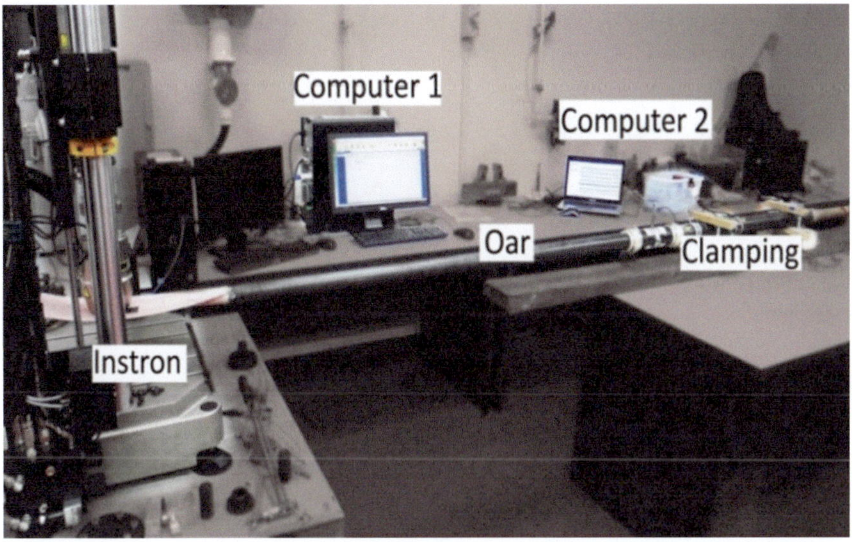

Fig. 7.17 Strain gauge calibration setup

displacement with a strain rate of 400 mm/min. The same setting was then applied to push the oar handle part. The change in strain for every force increase was then recorded using the computer.

The strain-force relationship obtained from the calibration work is shown in Fig. 7.18. Both of the graphs show a linear correlation as the force was applied. The gradient of the handle force graph is higher than the blade force since only small strains occur as the force is applied. Meanwhile, the handle shaft was bigger and shorter than the blade shaft, which was smaller in diameter and longer in length. The strain-force equations were then used to obtain the force exerted on the oar during the drive.

7.12 Sensor Accuracy and Repeatability Test

The accuracy of the measurement device is an important element to be considered before the experiment is carried out. Accuracy is defined as a measurement of the capability of the instrument device to faithfully indicate the value of the measurement compared to the actual measurement. This accuracy test was conducted on the encoder sensor to measure the oar angle and boat travelling distance before the data was converted into the speed value.

The setup of the mentioned accuracy test is shown in Fig. 7.19. The angle was measured between the two guide angles and the axis rotation of the oar. The guide angle was placed on the simplified rowing boat at zero and 45°. Initially, the oar

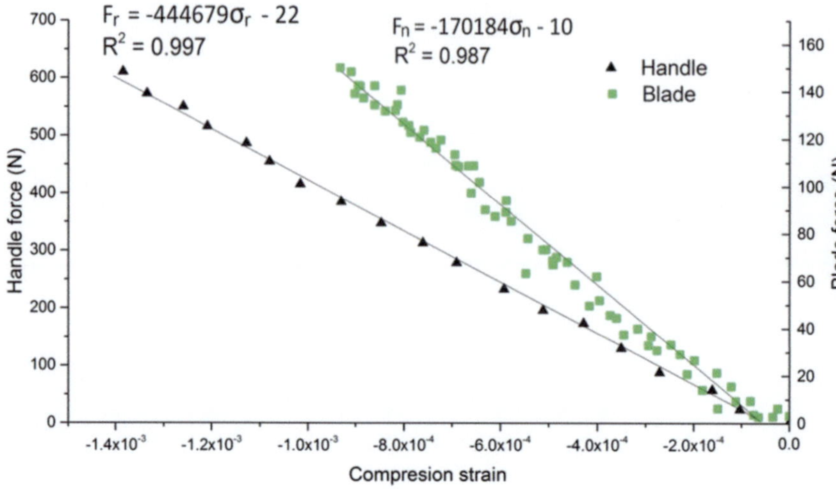

Fig. 7.18 The strain-force graph for the strain gauge calibration

was placed at zero-degree angles and then moved to 45°. The 45-degree angle was selected for the test as it was the maximum angle most rowers could reach in the hatch body position. The test was repeated five times, and the value was monitored using NI Signal Express software (National Instruments Corporation, Texas, USA). The average angle, standard deviation, and percentage error were calculated.

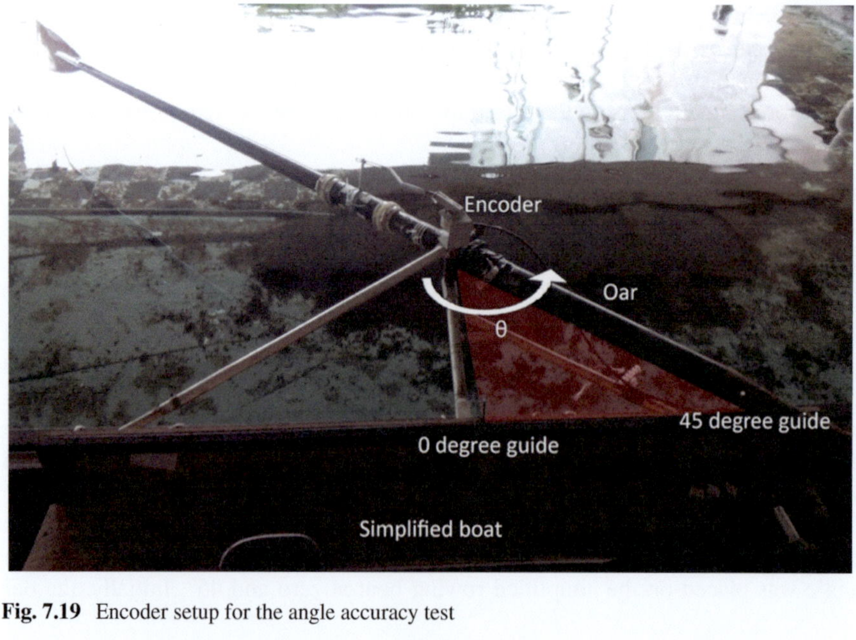

Fig. 7.19 Encoder setup for the angle accuracy test

The second accuracy test was carried out to measure the travelling distance read by the encoder along the rail. The test was carried out for 3 m distances. Zero distance and 3 m distance were marked on the rail. The simplified boat without a rower was then moved along the rail from zero distance to the 3 m distance. The test was repeated five times, and the data were collected using Signal Express software. The average travelling distance, standard deviation, and percentage error were calculated.

The test–retest reproducibility method used the Pearson correlation coefficient [20]. The test investigated the device's capability and the approach to obtain an identical value where a similar method was repeatedly measured. The test was carried out for ten rowers ($n = 10$). Selected rowers were asked to row on the rower simulator, and handle force was recorded. Handle force was selected as the important data for the reliability test because it powered up the rowing system. The result was then analysed using IBM SPSS Statistics 23 (IBM Corp, USA). The data was analysed based on Bivariate correlation using the Pearson correlation coefficient, where signification was set at 0.01 levels (2-tailed) [21].

The result of the first accuracy test was that the encoder sensor was accurate for measuring the angle. For a 45-degree measurement test with five times repetitions, the average result obtained was 45.44° with a standard deviation of 0.40. The average of the percentage error for the test was 1.05%. As for the measurement of the 3 m boat travelling distance, a 2.94 m average distance was obtained with a standard deviation of 0.023 m. The average percentage error of 1.903% was calculated. These percentage errors are acceptable since they are small and indicate good reading closeness to the measured value.

The second test of test–retest reliability analysis using ten rowers is shown in Table 7.3. Two data sets that repeated the same procedure are shown in columns 1 and 2.

Table 7.3 Data of the compression strain on the inboard shaft for test and re-test

Rower	Test 1 (Compression strain on the inboard shaft)	Test 2 (Compression strain on the inboard shaft)
1	– 0.00122	– 0.00121
2	– 0.00135	– 0.00134
3	– 0.00139	– 0.00132
4	– 0.00097	– 0.00099
5	– 0.00162	– 0.00169
6	– 0.00069	– 0.00069
7	– 0.00155	– 0.00155
8	– 0.00146	– 0.00148
9	– 0.00134	– 0.00132
10	– 0.00065	– 0.00065

Table 7.4 The Pearson correlation test result for test–retest ($n_r = 10$)

Test	Test 1	Test 2
Test 1	1	0.995
Test 2	0.995	1

Correlation is significant at the 0.01 level (2-tailed)

The Pearson correlation test result in Table 7.4 shows that for the group of the subject ($n_r = 10$), the test–retest correlation value of the measured handle force on the oar was 0.995 (at the significant level of $p < 0.05$). This indicates that the instrument and approach used are suitable for obtaining the repeatability data for the other subject in the next test [20].

7.13 Experimental Procedure

In the experiment, the boat moved from a stationary condition, which replicated the start of the race. The investigations began by pulling the oar handle towards the rower, as shown in Fig. 7.20. Consequently, the blade swept to generate the hydrodynamic force, and the rowing boat moved along the rail. The motion of the rower was captured using a camera (Sony FDR-AXP35) at 25 frames per second and was located 2 m from the boat rail, as shown in Fig. 7.21. It visualized the body segment movement during the stroke for the kinematic analysis work.

The oar angular velocity, boat linear velocity, rower handle force, and blade hydrodynamic force were recorded for further analysis. The first three strokes were tested at a 15 m travelling distance, and the test was repeated ten times. The experimental procedure simultaneously captured the handle force and blade hydrodynamics. Simultaneously analyzing these two aspects establishes the coupling mechanism between biomechanics and hydrodynamics of the rowing.

Fig. 7.20 Rower row during testing and oar blade positioned by submerging in water during the test

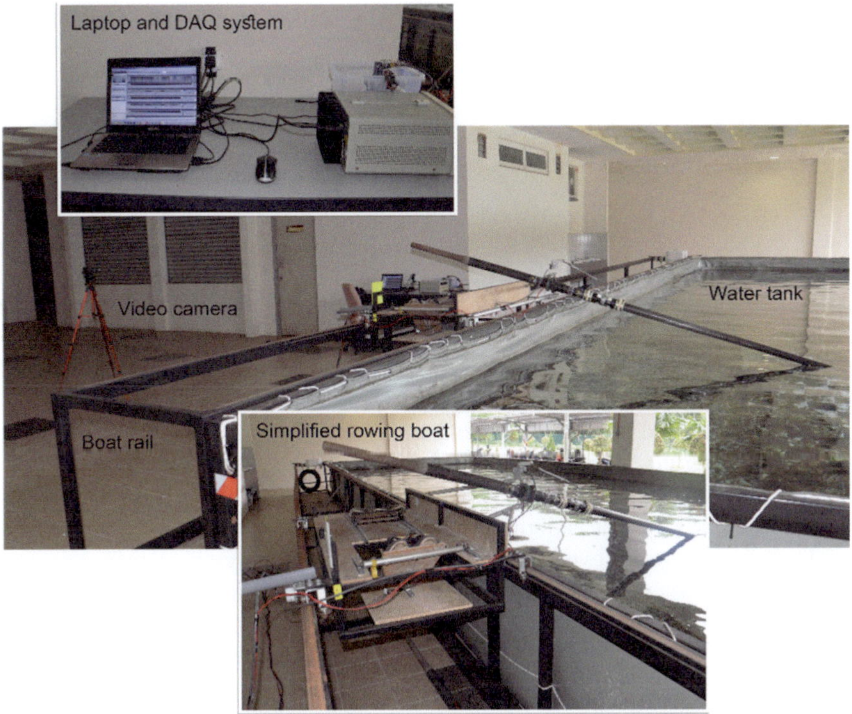

Fig. 7.21 Experimental setup for the study

7.14 Oar Efficiency

The oar is the intermediate component that transfers the force from the rower to the blade. It is important to evaluate the efficiency of the equipment that affects the rowing performance. The analysis efficiency is done under the following assumptions: (1) only small shaft bending occurs during the drive, (2) the submerged blade generates the hydrodynamic force, (3) air drag is ignored, (4) bearing friction is ignored, and (5) the boat moves in linear translation motion. The leverage efficiency of the oar is defined as the ratio of the power applied on the handle to the power of the propulsive force on the blade, as given in Eq. 7.3.

$$\eta_o = \frac{Power\ output\ at\ the\ blade}{Power\ input\ at\ the\ handle} = \frac{F_n d\dot{\theta}_c}{F_r l\dot{\theta}_c} \tag{7.7}$$

The power output is defined as the product of the blade's normal force, F_n, multiplied by the length of the load application point to the fulcrum location, d, with the oar angular velocity, $\dot{\theta}_c$ [22]. Whereas the power input is defined as the handle force, F_r, multiplied by the distance of the middle location handle part to the fulcrum location, l, with similar oar angular velocity, $\dot{\theta}_c$.

7.15 Kinematic of Rower

The Kinematic of the rower was captured using a video camera and analysed using Kinovea software, as shown in Fig. 7.22. The process started with setting up a camera at a suitable angle and distance to effectively capture the rower's movements. At the same time, the rower position was ensured to have good lighting conditions and a clear view. Since the simulator was outdoors, the white reflective stickers were used as the body markers. The reflective markers were used to identify key points on the rower's body and placed on the head, upper limb (wrist, elbow, and shoulders), pelvis, and lower limb (knee, ankle, and toe). The stroke session was captured from the side for better kinematic analysis, and the recorded session was imported into the Kinovea software. The calibration process is carried out on the video by setting a known distance between two points. This helps Kinovea convert pixel measurements to real-world measurements. Body marks were navigated from frame to frame to track its position. The distance and velocity of the body marks are measured automatically, as in Fig. 7.22.

7.16 Combination Between Biomechanics and Hydrodynamics of Rowing

The combination of the rower's biomechanics and the blade's hydrodynamics revealed the important aspect of energy generation and transfer involved in rowing, which had been simplified in other studies. The rowing power results from the rower's metabolic energy, which can be divided into the anaerobic metabolism (20–25% of energy) and the aerobic metabolism (75–80% of energy) [23]. This energy is then transformed into the kinetic energy of the rower through the speed and mass, as shown in Fig. 4.5. During the drive phase, the work given to the rowing system is positive, and the metabolic energy of the rower is released. During this phase, the work done benefits the rowing system by increasing the kinetic energy and accelerating the boat's speed.

The experimental results related to biomechanics provide valuable insights into rowing mechanics. This includes understanding how forces are applied to the oar, the movements of rowers, and the interaction between the rower and the boat. Such insights can lead to adjustments in training regimens, equipment design, and rowing techniques to optimize performance and reduce the risk of injury.

Meanwhile, during the recovery phase, the work done is negative. However, the metabolic energy is released during the return to the hatch body position, and the

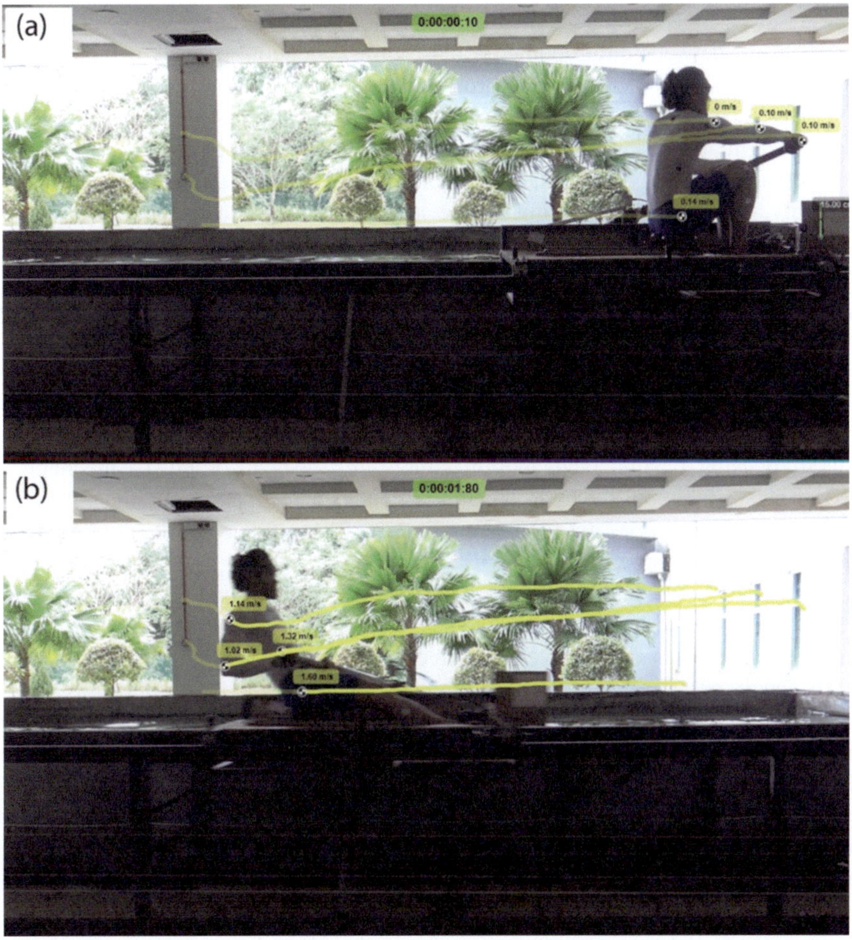

Fig. 7.22 Data captured using dynamic rowing simulator during stroke

boat is decelerated, as illustrated in Fig. 4.6. The rower's arms are an intermediate element for the energy transformation of the rotating blade. During the drive phase, the shaft stores some of the kinetic energy as the rebound energy through the shaft bending, besides acting as a damper to absorb the blade impact. Rowing propulsive occurred as the blade momentum imparted to the water, and this caused a momentum change, which generated the hydrodynamic force, as shown in Fig. 4.5.

The combination of biomechanics and hydrodynamics, emphasized in the current study, reveals the important aspect of the hydrodynamic force generated according to the rower biomechanics towards the rowing propulsive force. Most previous investigations ignored this aspect and the direct link to the boat's performance. R. M. Smith (2002) linked the stroke characteristic, the generated force, and the boat performance using the speed. It was found that the boat moved differently for different rowers due to the force generated by the rower [8]. Barret et al. investigated the relationship between anthropometry and rigging selection, rower kinematics, and physical capacity and evaluated this factor based on rowing performance [7]. M. M. Doyle et al. studied the biomechanics variable related to the force generated by the heavyweight and lightweight Men's coxless pair [22]. Although the assessment was correct, according to M. N. Macrossan et al., the only boat thrust comes from the blade hydrodynamic force during the acceleration. Thus, blade hydrodynamics must be considered to ensure a comprehensive study [24].

The hydrodynamics of rowing was evaluated based on the lift, drag, and hydrodynamic force generated under the influence of the rower biomechanics. N. Caplan et al. explored the blade's lift and drag coefficient pattern during the stroke to represent propulsive rowing [11]. It was discussed that the blade works similarly to a delta wing induced by the lift force. However, this study found that the oar blade works based on the drag induced dominated by the normal force. A. Sliasas et al. continued this study using dynamic blade motion, simplifying the blade propulsive mechanism [25]. The prediction of the hydrodynamic force is overestimated, and the pattern of the produced force does not represent either the biomechanics of the current study or the previous report.

A. Leroyer et al. studied using a rowing device [26]. Although the motion represented the boat and blade motion, the effect of the rower kinematics could not be included. A. Sliasas et al. studied the effect of the oar shaft on the propulsive force [27]. The graph report was similar to Fig. 4.5, where the first lever effect was incorporated. The hydrodynamics of the blade was lower due to the small oar mechanical advantage. However, the link to the boat performance was not included. The current study has improved those previous studies. It was found that the accuracy of the predicted hydrodynamic force was sensitive to the relative velocity of the blade and boat. Both values were related as the rower kinematics rotated the oar, and the blade generated the hydrodynamic force to accelerate the boat. Ignoring one of these important aspects may cause the rowing propulsive investigation incomplete.

In order to enhance rowing performance, the blade hydrodynamic force was important to be included in rowing investigation for performance evaluation or biomechanics investigation. The simplification of oar rotational speed and boat translation speed affects the blade's relative speed, leading to inaccuracy in the prediction of blade hydrodynamic force. A comparison between the current CFD and the study by Sliasas et al. [25] showed that the hydrodynamic peak force generated on the blade was 53% lower. This value was due to the rower powers the rowing propulsive and the consideration of the coupling mechanism between rower biomechanics and blade hydrodynamics, which was improved by the simplification approach used in the previous study.

References

1. Baudouin, A., Hawkins, D.: A biomechanical review of factors affecting rowing performance. Br. J. Sports Med. **36**(6), 396–402 (2002)
2. Sliasas, A., Tullis, S.: Numerical modelling of rowing blade hydrodynamics. Sports Eng. **12**(1), 31–40 (2009)
3. Baudouin, A., Hawkins, D.: Investigation of biomechanical factors affecting rowing performance. J. Biomech.Biomech. **37**(7), 969–976 (2004)
4. Soper, C., Hume, P.A.: Towards an ideal rowing technique for performance. Sports Med. **34**(12), 825–848 (2004)
5. Schwanitz, P.: Applying biomechanics to improve rowing performance. FISA Coach **2**(3), 1–7 (1991)
6. Kleshnev, V.: Blade propulsive efficiency; hydro-lift and drag forces on the blade. Rowing Biomech. Newslett. **7**(81) (2007)
7. Barrett, R.S., Manning, J.M.: Relationships between rigging set-up, anthropometry, physical capacity, rowing kinematics and rowing performance. Sports Biomech./Int. Soc. Biomech. Sports **3**(2), 221–235 (2004)
8. Smith, R.M., Loschner, C.: Biomechanics feedback for rowing. J. Sports Sci. **20**(10), 783–791 (2002)
9. Page, P.N., Hawkins, D.: A real-time biomechanical feedback system for training rowers. Sports Eng. **6**(2), 67–79 (2003)
10. Doyle, M.M., Lyttle, A., Elliott, B.: Comparison of force-related performance indicators between heavyweight and lightweight rowers. Sports Biomech. **9**(3), 178–192 (2010)
11. Caplan, N., Gardner, T.N.: Optimization of oar blade design for improved performance in rowing. J. Sports Sci. **25**(13), 1471–1478 (2007)
12. Meier, E.: The Wood Database. View (2013)
13. Smith, A., Wilkinson, S.J., Reynolds, W.N.: The elastic constants of some epoxy resins. J. Mater. Sci. **9**(4), 547–550 (1974)
14. Ibtihal-Al-Namie, Study the mechanical properties of epoxy resin reinforced with silica (quartz) and alumina particles. Iraqi J. Mech. Mater. Eng. **11**(3) (2011)
15. Rahmani, H., Mahmoudi, S.H., Ashori, A.: Mechanical performance of epoxy/carbon fiber laminated composites. J. Reinforced Plastics Compos. (2014). https://doi.org/10.1177/0731684413518255
16. Micro-Measurements, V.: The three-wire quarter-bridge circuit. Appl. Note TT **612**, 221–223 (2010)
17. Laschowski, B., et al.: The effects of oar-shaft stiffness and length on rowing biomechanics. Proc. Inst. Mech. Eng. Part P: J. Sports Eng. Technol. (2015). https://doi.org/10.1177/1754337115582121
18. Kleshnev, V.: Points of force application to the oar and efficiency of various blade designs (2003). www.biorow.com/Papers_files/
19. Kleshnev, V.: Boat acceleration, temporal structure of the stroke cycle, and effectiveness in rowing. Proc. Inst. Mech. Eng. Part P: J. Sports Eng. Technol. **224**(1), 63–74 (2010)
20. Piaw, C.Y.: Mastering Research Statistics. McGraw Hill Education, Shah Alam, Malaysia (2013)
21. Wunderlich, M., et al.: A new approach to assess movements and isometric postures of spine and trunk at the workplace. Eur. Spine J. **20**(8), 1393–1402 (2011)
22. Machado, L., et al.: The effect of depth on the drag force during underwater gliding: a CFD approach. In: ISBS-Conference Proceedings Archive (2010)
23. Nilsen, T.S., Daigneault, T., Smith, M.: Basic Rowing Physiology. FISA Development Commission (1990)
24. Macrossan, M.N., Macrossan, N.W.: Energy Efficiency of the Rowing Oar from Catch to Square-off (2008)

25. Sliasas, A., Tullis, S.: The dynamic flow behaviour of an oar blade in motion using a hydrodynamics-based shell-velocity-coupled model of a rowing stroke. Proc. Inst. Mech. Eng. Part P: J. Sports Eng. Technol. **224**(1), 9–24 (2010)
26. Leroyer, A., et al.: Experimental and numerical investigations of the flow around an oar blade. J. Mar. Sci. Technol. **13**(1), 1–15 (2008)
27. Sliasas, A., Tullis, S.: Modelling the effect of oar shaft bending during the rowing stroke. Proc. Inst. Mech. Eng. Part P: J. Sports Eng. Technol. **225**(4), 265–270 (2011)

Chapter 8
Conclusion and Recommendation

In general, this book highlights the relevant information related to rowing. The current state of rowing research focuses on rowing performance primarily associated with the biomechanics of rowing directly to the boat performance without considering the hydrodynamics of the blade. Simultaneously, the hydrodynamic of the blade is studied using the simplified rowing mechanism.

There are several research gaps identified. The first research gap is limited studies that directly analyze coupling mechanisms between rower biomechanics and blade hydrodynamics. The second research gap is the kinematics of the rower during the stroke, which was not included in the study. The third research gap is the hydrodynamic of the blade due to the rower kinematics; power stroke is not being considered. Finally, the fourth research gap is the hydrodynamic aspect of the blade design, and features are limited and not explored in detail.

The oar blade is essential to generate propulsive force and drive the boat forward. The analysis of the fluid flow around the blade was reported by the previous study using computational and simplified experiment tests to collect data related to important oar blade parameters. It is suggested for better blade performance:

1. The blade is designed to cover a more extensive area with a more significant aspect ratio.
2. The curvature of the blade is considered at the width axis.
3. The widest section is in the middle of the blade length.
4. Position the blade horizontally parallel to the water surface.

This study suggests the coupling mechanism between rower biomechanics and blade hydrodynamics for comprehensive rowing performance investigation. By mimicking the actual mechanism of the rowing boat, a rowing simulator can be used to understand the rowing mechanism and how forces were transmitted from rower biomechanics to blade hydrodynamics. Based on the lever's mechanical advantage, the corresponding blade hydrodynamics force is lower than the handle force. This force reduction was due to the distance between the applied loads to the rotational

© The Author(s), under exclusive license to Springer Nature Singapore Pte Ltd. 2025
A. A. Mohd Yusof et al., *Rowing Biomechanics and Hydrodynamics*,
Series in BioEngineering, https://doi.org/10.1007/978-981-97-7271-1_8

Table 8.1 Future recommendation for rowing study

Future recommendations	Detail
Further understanding of kinematic effects of rower to rowing performance	Investigate the influence of rower kinematics on blade hydrodynamics. Understanding how the rower's movements affect the blade's performance can provide valuable insights for optimizing rowing techniques and equipment. Develop a comprehensive model integrating rower kinematics and blade hydrodynamics to understand the rowing propulsive force better. This could help bridge the gap between biomechanics and boat performance.
Biomechanics-performance link	Investigate the direct link between rower biomechanics and boat performance. How do different biomechanical factors contribute to boat speed, and how can rowers optimize their technique based on this knowledge?
Individual differences	Examine how individual differences in rowers, such as anthropometry, physical capacity, and skill level, affect the biomechanics and hydrodynamics of rowing. This can help tailor training and equipment recommendations for different athletes
Computational fluid dynamics (CFD)	Further research can use advanced CFD techniques to provide more accurate predictions of hydrodynamic forces during rowing. This could involve simulating the interaction between the blade and water in more complex and realistic ways
Oar design and mechanical advantage	Investigate how variations in oar design, including the oar shaft and blade shape, impact rowing performance and the generation of hydrodynamic forces. This could lead to recommendations for optimizing our design

axis. It is suggested that to counter the force reduction, the oar blade should focus on improving the blade efficiency to make the blade act as an anchor underwater support condition.

Understanding the interaction between rower biomechanics and blade hydrodynamics is essential for advancing rowing techniques. As rowers exert energy to propel their oars, it's vital to comprehend how this energy translates into boat propulsion. Integrating biomechanical and hydrodynamic insights reveals the complicated relationship between rower movements and oar performance, directly impacting boat speed (Table 8.1).

MIX
Papier aus verantwortungsvollen Quellen
Paper from responsible sources
FSC® C105338
FSC
www.fsc.org

If you have any concerns about our products,
you can contact us on
ProductSafety@springernature.com

In case Publisher is established outside the EU,
the EU authorized representative is:
Springer Nature Customer Service Center GmbH
Europaplatz 3, 69115 Heidelberg, Germany

Printed by Libri Plureos GmbH
in Hamburg, Germany